FROM THE BRAIN TO THE MOUTH

NEUROPSYCHOLOGY AND COGNITION

VOLUME 12

The purpose of the Neuropsychology and Cognition series is to bring out volumes that promote understanding in topics relating brain and behavior. It is intended for use by both clinicians and research scientists in the fields of neuropsychology, cognitive psychology, psycholinguistics, speech and hearing, as well as education. Examples of topics to be covered in the series would relate to memory, language acquisition and breakdown, reading, attention, developing and aging brain. By addressing the theoretical, empirical, and applied aspects of brain-behavior relationships, this series will try to present the information in the fields of neuropsychology and cognition in a coherent manner.

The titles published in this series are listed at the end of this volume.

From the Brain to the Mouth

Acquired Dysarthria and Dysfluency in Adults

Edited by

Yvan Lebrun
Vrije Universiteit Brussel / Université Libre de Bruxelles

KLUWER ACADEMIC PUBLISHERS
DORDRECHT / BOSTON / LONDON

A C.I.P. Catalogue record for this book is available from the Library of Congress.

ISBN 0-7923-4427-8

Published by Kluwer Academic Publishers,
P.O. Box 17, 3300 AA Dordrecht, The Netherlands.

Kluwer Academic Publishers incorporates
the publishing programmes of
D. Reidel, Martinus Nijhoff, Dr W. Junk and MTP Press.

Sold and distributed in the U.S.A. and Canada
by Kluwer Academic Publishers,
101 Philip Drive, Norwell, MA 02061, U.S.A.

In all other countries, sold and distributed
by Kluwer Academic Publishers Group,
P.O. Box 322, 3300 AH Dordrecht, The Netherlands.

Printed on acid-free paper

Printed in the Netherlands

TABLE OF CONTENTS

PREFACE

The impetus for the present volume was provided by a European conference held at St. Martens Latem (Belgium) in September 1994 and sponsered by the Directorate General XII (Science, Research and Development) of the European Commission.

Preparation of the volume was also financially supported by the Directorate General.

The editor, the contributors and all the participants in the conference are grateful to the European Commission for the generous help received.

CONTRIBUTORS

Hermann ACKERMANN, Department of Neurology, University of Tübingen, 3 Hoppe-Seyler-Straße, 72076 Tübingen (Germany).

Henny BIJLEVELD, Philologie Germanique, Faculté de Philosophie et Lettres, Université Libre de Bruxelles, 50 av. F. Roosevelt, 1050 Brussels (Belgium).

Veronique CAPPAERT, University Hospital Ghent, 2P1 Centrum voor Gehoor- en Spraakrevalidatie, 185 De Pintelaan, 9000 Gent (Belgium).

Patrick COPPENS, Speech-Language-Hearing Sciences, Moorhead State University, Moorhead, MN 56563 (USA).

Jeanine DELEVAL, Department of Neurology, University Hospital 'Erasme', 808 Route de Lennik, 1070 Brussels (Belgium).

Christine EECKHAUT, University Hospital Ghent, 2P1 Centrum voor Gehoor- en Spraakrevalidatie, 185 De Pintelaan, 9000 Gent (Belgium).

Ingo HERTRICH, Department of Neurology, University of Tübingen, 3 Hoppe-Seyler-Straße, 72076 Tübingen (Germany).

Margaret LEAHY, Clinical Speech and Language Studies, Trinity College, Dublin 2 (Ireland).

Yvan LEBRUN, Neurolinguistics Department, School of Medicine VUB, 103 Laarbeeklaan, 1090 Brussels (Belgium).

Nadine MOREAU, Laboratoire de Neuropsycholinguistique Jacques Lordat, Département des Sciences du Langage, Université de Toulouse-Le Mirail, 5 allée A. Machado, 31058 Toulouse Cedex (France).

Jean-Luc NESPOULOUS, Laboratoire de Neuropsycholinguistique Jacques Lordat, Département des Sciences du Langage, Université de Toulouse-Le Mirail, 5 allée A. Machado, 31058 Toulouse Cedex (France).

Kristien OOSTRA, University Hospital Ghent, 2P1 Centrum voor Gehoor- en Spraakrevalidatie, 185 De Pintelaan, 9000 Gent (Belgium)

Philippe PAQUIER, Department of Neurology, University Hospital 'Erasme', 808 Route de Lennik, 1070 Brussels (Belgium), and Unit of Applied Neurolinguistics, School of Medicine UIA, Antwerpen (Belgium).

Janneke POOT, Department of Neurology, EE 2287, University Hospital Dijkzigt, Molewaterplein 40, 3015 GD Rotterdam (The Netherlands).

Randall ROBEY, Communication Disorders Program, University of Virginia, 2205 Fontaine Avenue - Suite 202, Charlottesville, VA 22903 (USA).

Anne-Marie SIMON, Laboratoire de Recherches sur le Langage et les Handicaps de la Communication, Hôpital de la Salpêtrière, Boulevard de l'Hôpital, 75013 Paris (France).

Trudy STEWART, Speech and Language Therapy Department, St. James's University Hospital, Leeds (U.K.).

Hugo van DONGEN, Department of Neurology, EE 2287, University Hospital Dijkzigt, Molewaterplein 40, 3015 GD Rotterdam (The Netherlands).

Pieter van DOORN, Department of Neurology, EE 2287, University Hospital Dijkzigt, Molewaterplein 40, 3015 GD Rotterdam (The Netherlands).

Kristiane van LIERDE, University Hospital Ghent, 2P1 Centrum voor Gehoor- en Spraakrevalidatie, 185 De Pintelaan, 9000 Gent (Belgium).

Manfredi VENTURA, Department of Neurology, University Hospital 'Erasme', 808 Route de Lennik, 1070 Brussels (Belgium).

John VAN BORSEL, University Hospital Ghent, 2P1 Centrum voor Gehoor- en Spraakrevalidatie, 185 De Pintelaan, 9000 Gent (Belgium).

Ellen YOUSEF-BAK, Department of Neurology, EE 2287, University Hospital Dijkzigt, Molewaterplein 40, 3015 GD Rotterdam (The Netherlands).

INTRODUCTION

Y. LEBRUN

ACQUIRED DYSARTHRIA AND DYSFLUENCY IN ADULTS

In comparison with aphasia, which has been extensively studied during the last 130 years, dysarthria has attracted relatively little attention. Many more publications have been devoted to disorders of language than to disorders of motor speech. Indeed, dysarthria has often been treated as a Cinderella by language pathologists. Hécaen and Dubois (1971), for instance, excluded motor speech impairments from the field of neurolinguistic research.

Those who did deal with dysarthria frequently contented themselves with impressionistic descriptions which, though they were sometimes couched in colourful terms, did little justice to the true nature of the condition. When Jelgersma (quoted by Lhermitte, 1958, p. 444) stated that the patient with cerebellar dysarthria tended to chew his words before spitting them out, he added nothing to our understanding of the disturbance. Indeed it may even be doubted whether his description was a fair rendition of the verbal behaviour of patients with cerebellar dysarthria.

Fortunately things have started to change. Disorders affecting the programming and organisation of phono-articulatory movements attract more and more attention, and cursory impressionistic descriptions are progressively replaced by accurate accounts based on careful examination and objective measurements.

Yet, many problems remain to be solved. Despite undeniable progress in the analysis and characterization of the various types of dysarthria, a large number of questions are still unanswered. New clinical and experimental evidence as well as novel and stimulating views are necessary to improve our knowledge of the pathophysiology of motor speech disorders. The purpose of the present book is precisely to report fresh observations and to present provocative thoughts so as to further the understanding of dysarthria.

Dysarthria may be broadly defined as the inability to correctly form spoken words and/or to concatenate them fluently. In dysarthric patients pronunciation is distorted and/or delivery is disrupted. In some patients only the first disorder is in evidence, in others, only the second. In still others, the two impairments are simultaneously present.

Not infrequently, the word *dysarthria* is used restrictively to refer to the first disturbance only (distorted pronunciation) and the word *dysfluency* is applied the second (disrupted delivery). This usage has been adopted in the present work.

This should not obliterate the fact that dysarthria and dysfluency represent two sides of the same coin. Both of them are motoric disorders of

Y. Lebrun (ed.), From the Brain to the Mouth, 3–8.

speech. They are disturbances of oroverbal performance. In the present book the close relationship between the two impairments is emphasized, and it is argued that some cerebral structures, such as the extrapyramidal system, can be implicated in either of them.

It is generally agreed that dysarthria in the restricted sense of the word results from brain damage, even though the possibility of there being a functional or psychogenic dysarthria is sometimes entertained, as the contribution by VAN DONGEN and colleagues shows. However, there appears to be no clear indication that acquired dysarthria in adults is ever psychologically based.

Acquired dysarthria in adults is a neuromuscular disorder: the patient knows what words to use but is unable to command his speech organs so as to utter the intended words without distortion. This inability may result from damage to various components of the central nervous system. The dysarthric deviations tend to vary with the localisation of the causal lesion(s).

In their contribution, HERTRICH and ACKERMANN provide a wealth of quantitative information on the temporal organization of syllabic and intrasyllabic segments in four types of dysarthria. Their measurements can be used as a touchstone for a number of statements regarding speech rate in dysarthric patients.

Rate of delivery in patients with extrapyramidal diseases is still a matter of debate. It is generally agreed that affections of the extrapyramidal system may entail precipitate speech. But it is not clear how frequent tachyphemia actually is in patients with lesions in the basal ganglia. Séguier et al. (1974) found that as many as 38% of the Parkinsonian patients they examined evidenced abnormally fast verbal delivery. However, speed of delivery was judged auditorily, and auditory perceptive impressions may be quite misleading. Speech may sound precipitate whilst actually articulatory velocity is not augmented. Palilalic speech offers a good example of this. It has repeatedly been stated that in palilalias there is a progressive increase in speech rate. In Boller and co-workers' words (1973), "palilalia is a disorder of speech characterized by compulsive repetiton of a phrase or word that the patient reiterates with increasing rapidity." However, acoustic measurements by Kent and LaPointe (1982) and by Ackermann et al. (1989) have shown that this is not always the case: the final repetition(s) in a palilalia may last as long as, if not longer than, the preceding ones.

This example shows how crucial objective measurements can be for the correct appraisal of dysarthric deviances. Accordingly, the data provided by HERTRICH and ACKERMANN in this volume should prove very useful in future research. In addition, their contrastive findings suggest interesting hypotheses regarding dyscoordination of the speech organs in patients with extrapyramidal diseases as opposed to patients with cerebellar affections. These hypotheses open new vistas for further investigation.

Demyelinating lesions in the cerebral white matter may entail speech and language disorders. The former may include hypokinetic articulation as well as a tendency to repeat one's words (Lebrun et al., 1991). In the case reported by COPPENS and ROBEY in this volume there was dysarthria without concomitant aphasia. Careful analysis of the patient's motor speech disorder revealed, however, that not all of the dysarthric features could be ascribed to the demyelinating lesions in the hemispheric white matter. A second disease had to be assumed that affected a lower part of the neuraxis. Subsequent examination verified the hypothesis. This case, then, shows how useful a detailed analysis of dysarthria can be. Such analysis may uncover an unsuspected neurological affection.

Ever since Pierre Marie (1906a, b) at the turn of the century coined the formula Broca's aphasia = (Wernicke's) aphasia + anarthria, aphasiologists have wondered about the exact nature of the segmental distortions observed in the oral output of motor aphasics. Are these distortions dysarthric and is aphasia thus concomitant with dysarthria in these patients, or are they qualitatively different from the phono-articulatory deviations that occur in the speech of patients with neuromuscular problems? The contribution of NESPOULOUS and MOREAU should help shed light on this issue. Re-analyzing previously published data, they show that Broca's aphasics may both destroy and create consonantal clusters. Moreover, clusters may be destroyed either by vocalic epenthesis or consonantal syncope. The latter does not occur haphazardly but affects certain consonants preferentially. Further research should reveal whether a similar pattern is to be found in dysarthria, thus supporting Marie's view that Broca's aphasia is actually a compound disorder.

Pierre Marie never succeeded in clearly delineating the syndrome he called *anarthria* (Lebrun and Martinez, 1991; Lebrun 1994). The term, therefore, is used to refer to various entities. Some apply it to extreme forms of dysarthria, when motor speech has completely broken down and the patients are totally unable to form words orally. For instance, Vallar and Cappa (1987) spoke of anarthria in the case of a locked-in syndrome patient who was unable to move and accordingly to speak, but who had preserved his linguistic competence. Anarthria, then, is sometimes used as the superlative of dysarthria.

Other clinicians, including VENTURA, PAQUIER and DELEVAL in

this volume, prefer to apply the name to apraxia of speech. The latter is a controversial syndrome. Some speech pathologists ignore it altogether or even deny its existence. Others consider that apraxia of speech exists really but occurs rarely in isolation; most of the time it is observed in combination with motor (or Broca's) aphasia. Still others believe that apraxia of speech is a syndrome in its own right, qualitatively different from the output difficulties of motor (or Broca's) aphasics (Lebrun, 1989, 1990).

At times *aphemia* is used instead of *anarthria* to denote apraxia of speech. In their contribution to this volume VENTURA and his co-workers propose to distinguish between aphemia, on the one hand, and apraxia of speech or anarthria, on the other hand. In their view, aphemia can be conveniently used to refer to sudden and complete speechlessness following acute brain damage. The patient evidences a desire to communicate but is completely unable to speak. Yet, comprehension of both oral and written language is largely preserved and so is writing. This condition is most of the time transitory. After a while, depending on the precise location of the lesion within the frontal lobe, it evolves into speech apraxia, dysarthria or motor aphasia.

While some adults following brain damage lose the ability to speak, others become unable to control their delivery: they block on a number of sounds or involuntarily repeat speech segments. Their dysfluency resembles stuttering as it is observed in a number of children. It is therefore often called adult-onset stuttering.

Just as dysarthria has been neglected in comparison with aphasia, so has adult-onset stuttering relatively to developmental stuttering. While the latter has been extensively studied both clinically and experimentally, and has been made the subject-matter of innumerable publications in the last 15O years, it is only of late that acquired stuttering in adults started to be looked at with more than a cursory glance. Thanks to recent research, the various forms of adult-onset stuttering begin to be better known and understood. Yet, many questions remain.

Foremost among the debated issues is the nature of adult-onset stuttering. Is this disorder neurogenic or psychogenic? And if it is at times organically and at times psychologically based, how is differential diagnosis to be made? LEBRUN and LEAHY and STEWART tackle this problem in the present volume. At first sight they may seem to take divergent views. In reality LEBRUN's organic interpretation leaves room for the affective components emphasized by LEAHY and STEWART.

BIJLEVELD and SIMON report a case of adult-onset stuttering whose organicity can hardly be doubted, whereas VAN BORSEL, VAN LIERDE,

OOSTRA and EECKHAUT describe a teenager with brain trauma whose acquired dysfluency seems to have been reactive. These two cases are particularly interesting when read in the light of the controversy discussed by LEBRUN and by LEAHY and STEWART. In addition, BIJLEVELD and SIMON's findings suggest a link between oral-verbal fluency and manual agility. This link fits in well with the pathogenetic views put forward by LEBRUN at the end of his contribution.

In a number of individuals it is observed that developmental stuttering abates or disappears with time but recurs in adulthood following brain damage. The chapter by VAN BORSEL and CAPPART deals with this particular type of dysfuency and considers a number of hypotheses regarding its pathophysiology. Their contribution nicely supplements the chapters in the volume that are devoted to acquired stuttering in adults.

The ten chapters in the present book, then, confirm the theoretical and clinical relevance of the analysis of motor speech disorders and present a number of observations and of considerations likely to improve the understanding of the disturbances which may affect the programmation and performance of phono-articulatory gestures.

REFERENCES

ACKERMANN H., ZIEGLER W., OERTEL W. (1989) Palilalia as a symptom of levodopa induced hyperkinesia in Parkinson's disease. *Journal of Neurology, Neurosurgery and Psychiatry* 52: 805-807.

BOLLER F., BOLLER M., DENES G., TIMBERLAKE W., ZIEPER I., ALBERT M. (1973) Familial palilalia. *Neurology* 23: 1117-1125.

HECAEN H., DUBOIS J. (1971) La neurolinguistique. In PERREN G., TRIM J. (eds.) *Application of linguistics*. London, Cambridge University Press: 85-99.

KENT R., LAPOINTE L. (1982) Acoustic properties of pathologic reiterative utterances: A case study of palilalia. *Journal of Speech and Hearing Research* 25: 95-99.

LEBRUN Y. (1989) Apraxia of speech: The history of a concept. In SQUARE-STORER P. (ed.) *Acquired apraxia of speech in aphasic adults*. London, Taylor and Francis: 3-19.

LEBRUN Y. (1990) Apraxia of speech: A critical review. *Journal of Neurolinguistics* 5: 379-406.

LEBRUN Y. (1994) Pierre Marie. In ELING P. (ed.) *Reader in the history of aphasia*. Amsterdam, Benjamins: 219-250.

LEBRUN Y., DEVREUX F., ROUSSEAU J. (1991) Language and speech in a case of perivenous encephalomyelitis. *Aphasiology* 5: 87-95.

LEBRUN Y., MARTINEZ C. (1991) L'anarthrie de Pierre Marie. *Glossa* 23: 12-15.

LHERMITTE F. (1958) Le syndrome cérébelleux. Etude anatomo-clinique chez l'adulte. *Revue Neurologique* 98: 435-477.

MARIE P. (1906a) La troisiôme circonvolution frontale ne joue aucun rêle spécial dans la fonction du langage. *La Semaine Médicale* 26: 241-247.

MARIE P. (1906b) Que faut-il penser des aphasies sous-corticales (aphasies pures)? *La Semaine Médicale* 26: 493-500.

SEGUIER N. SPIRA A., DORDAIN M., LAZAR P. CHEVRIE-MULLER C. (1974) Etude des relations entre les troubles de la parole et les autres manifestations cliniques dans la maladie de Parkinson. *Folia Phoniatrica* 26: 108-126.

VALLAR G., CAPPA S. (1987) Articulation and verbal short-term memory: Evidence from anarthria. *Cognitive Neuropsychology* 4: 55-78.

DYSARTHRIA

I. HERTRICH, H. ACKERMANN

ACOUSTIC ANALYSIS OF DURATIONAL SPEECH PARAMETERS IN NEUROLOGICAL DYSARTHRIAS

1. INTRODUCTION

Darley, Aronson and Brown (1969a,b, 1975) performed the only comprehensive investigation of neurological dysarthrias available so far. About eight of the 38 perceptual dimensions of dysarthric speech considered by these authors refer to the temporal structure of verbal utterances, e.g. "rate", "excess and equal stress", "prolonged phonemes". Disturbances in this domain even belong to the most prominent features of choreic and ataxic dysarthria.

Auditory assessment of speech disorders, however, is not a sufficiently reliable procedure (Zyski and Weisiger 1987). Further elucidation of the disordered temporal organization of speech utterances in neurological dysarthrias, therefore, requires parametric data. Acoustic analyses of the speech signal provide objective and quantitative measures in these regards. Patients with Parkinson's disease often present with normal speech tempo in terms of syllable, word and sentence lengths (Ludlow et al. 1984; Ludlow et al. 1987; Ackermann and Ziegler 1991). However, there seems to be a rather small subgroup of Parkinsonians with reduced speaking rate. Measurements at the subsyllabic level yielded discrepant results including both segmental lengthenings and shortenings (Weismer 1984). During rapid syllable repetitions (oral diadochokinesis tasks) reduced, normal as well as gradually increasing rates were observed (Kreul 1972; Hirose et al. 1981; Ludlow et al. 1987). Besides Parkinsonism Huntington's disease represents another paradigm of basal ganglia dysfunctions. When asked to speak at a convenient and at a fast rate, respectively, these subjects showed less changes in sentence duration than their controls (Ludlow et al. 1987). Accordingly, reduced maximum speech rate has been measured in this group (Volkmann et al. 1992). Moreover, Hertrich and Ackermann (1994) reported increased variability of utterance duration and/or voice-onset-time (VOT) across sentence repetitions even in non-dysarthric patients with Huntington's disease. Acoustic studies noted slowed speech and reduced syllable repetition rates in ataxic subjects (Kent et al. 1979; Portnoy and Aronson 1982; Gentil 1990a,b; Ackermann and Hertrich 1993). Increased syllable lengths during sentence utterances were measured in

Y. Lebrun (ed.), From the Brain to the Mouth, 11–47.

spastic dysarthria due to bilateral lesions of the corticobulbar tracts (Ziegler and von Cramon 1986).

The investigations referred to used different speech materials and measurement procedures. Thus, the results provided by these various studies cannot be unambiguously compared with each other. Moreover, only sparse data about subsyllabic segments, e.g. the duration of vowels and VOT, are available so far. It is not possible, thus, to infer on these grounds comprehensive differential-diagnostic patterns of speech timing in neurological dysarthrias.

In order to provide a detailed acoustic analysis of the temporal organization of syllabic and intrasyllabic segments in various central-motor disorders, well-documented patients suffering from idiopathic Parkinson's disease (PD), Huntington's disease (HD), Friedreich's ataxia (FA), and purely cerebellar ataxia (CA) were studied using identical tests and parameters in all subjects. PD and HD represent dysfunctions of the basal ganglia, FA mainly reflects a disorder of the cerebellar afferents. With the exception of a single subject all patients with cerebellar ataxia showed an atrophy restricted to the cerebellar cortex. Thus, disorders of three distinct components of the cerebral motor system, i.e. basal ganglia, cerebellar afferents and cerebellar cortex, were considered.

2. GENERAL METHODS

2.1 *Subjects*

Table 1 presents the relevant clinical data of the four patient groups examined. Overall performance during spontaneous speech of the PD, HD, and FA subjects, as evaluated by auditory-perceptual ratings, extended from unimpaired verbal communication to severe dysarthria in terms of mostly unintelligible utterances. In contrast, the CA patients were, as a whole, less impaired. The most disabled individuals in this group showed moderately reduced intelligibility only.

With one exception (HD7) all HD subjects had a definite family history of their disease (Table 1A). Genealogic documentation of patient HD7 was incomplete. In this case diagnosis relied on the typical medical history with progressive development of mental alterations, emotional disturbances, and choreic hyperkinesia. Furthermore, clinical and laboratory data did not provide any indications of a symptomatic variant of this disorder. All but one patient presented with chorea at the time of speech examination. In more advanced cases of HD an akinetic-rigid syndrome may replace hyperkinesia. This development had occurred in patient HD10. Neuropsychological testing disclosed slight to moderate cognitive and memory deficits in most subjects. Nevertheless, all HD individuals were fully able to cooperate and none was disoriented to time and place. Eleven patients showed reduced volume of the caudate nucleus at cranial computerized tomography and/or magnetic resonance imaging scans. Missing caudate atrophy in subjects HD1 and HD3 might have been due to short disease duration. Patient HD2 refused neuroradiological examination.

Seventeen subjects had a diagnosis of idiopathic PD on the grounds of history and clinical findings (Table 1B). Apart from a single individual (PD2) all PD patients were receiving dopaminergic medication at the time of speech investigation. Subjects with motor fluctuations were examined during an akinetic state. Cranial computerized tomography and/or magnetic resonance imaging revealed slight or moderate cerebral atrophy in some patients. Severe atrophy was considered a criterion of exclusion from the present study.

Eleven FA subjects were tested (Table 1C). All but four patients presented with a family history of FA. With respect to cranial nerve functions a peripheral-vestibular deficit was disclosed in a single patient (FA4). A further subject (FA10) showed synkinesia of facial muscles reflecting pathological reinnervation.

On the grounds of history, clinical findings, and neuroradiological data a purely cerebellar syndrome (CA) was assumed in 14 patients (Table 1D). Eleven subjects out of this group (CA1–11) presented with a slowly progressive course. In consideration of their family history four of them (CA1,8,9,10) had a diagnosis of autosomal dominant cerebellar ataxia (ADCA III, Harding 1984). The remaining seven cases with chronic ataxia were identified as idiopathic cerebellar atrophy. Three patients of the CA group suffered from an ataxic syndrome of acute onset due to cerebellitis (CA12,14) or lithium intoxication (CA13). CCT and/or MRI scans revealed atrophy of the cerebellum in all but one (CA12) CA subject. Significantly reduced volume of the brainstem was considered a criterion of exclusion from this study.

The control group included 15 subjects (9 men, age 26–72 years; 6 women, age 31–70 years) who had never suffered from a disease or trauma affecting the central nervous system or the cranial nerves.

2.2 *Test materials*

All subjects underwent an articulation test comprising single-word utterances, sentence repetitions, isolated vowels, oral diadochokinesis tasks,

TABLE 1A
Clinical data of the patients suffering from Huntington's disease (HD)

Pat	Age	Sex	Dur	Family history	Stage	Cogn dysf	Motor signs	Caudate atrophy	Medication
HD1	30	f	<1	+	1	+	chorea	-	-
HD2	33	m	<1	+	1	-	chorea	?	-
HD3	35	f	1	+	3	+	chorea	-	sulp
HD4	56	f	2	+	2	+	chorea	+	-
HD5	42	m	2	+	3	+	chorea	+	-
HD6	54	m	3	+	2–3	+	chorea	+	sulp
HD7	64	f	3	?	2	+	chorea	+	sulp
HD8	51	m	5	+	2	+	chorea	+	sulp
HD9	33	m	7	+	4	+	chorea	+	-
HD10	34	m	8	+	4	+	aki/rig	+	-
HD11	49	m	8	+	4	+	chorea	+	sulp
HD12	57	m	8	+	4	+	chorea	+	sulp
HD13	53	f	10	+	2–3	+	chorea	+	-
HD14	57	m	10	+	4	+	chorea	+	sulp

Age: in years; Dur = disease duration (years); Stage: according to Shoulson and Fahn (1979); Cogn dysf = cognitive dysfunction; aki/rig = akinetic-rigid syndrome; Caudate atrophy = atrophy of the caudate nucleus; sulp = sulpiride; + = present; - = absent.

TABLE 1B
Clinical data of the patients suffering from Parkinson's disease (PD)

Pat	Age	Sex	Dur	Motor signs (extremities)			Depression	Dementia	Cerebral atrophy	Medication
				Akinesia	Rigidity	Tremor				
PD1	69	f	1	2	1	-	-	-	-	ld
PD2	71	m	2	2	1	2	-	-	-	ac
PD3	74	m	3	2	2	1	1	-	2	ld
PD4	60	f	4	2	2	-	-	-	?	ld,br,am,se
PD5	60	m	4	1	-	2	-	-	1	ld,ac
PD6	58	m	5	1	-	2	-	-	-	ld,br,se
PD7	63	m	5	1	1	-	1	-	-	ld,am
PD8	52	f	8	2	1	-	-	-	-	ld,br
PD9	72	m	8	2–3	2	1	-	-	2	ld,br,am,se
PD10	57	m	9	2	2	1	1	-	1	ld,br
PD11	58	f	10	1	1	1	-	-	-	ld,br,ac
PD12	54	m	11	2	1	-	-	-	-	ld,br,se
PD13	60	m	11	2	1	-	1	-	1	ld,br,se
PD14	47	m	12	1	-	-	-	-	-	ld,am
PD15	66	m	12	2	1	-	-	-	-	ld,br
PD16	76	m	13	3	2	2	1	-	1	ld,br
PD17	66	m	30	2-3	2	-	-	-	-	ld,am,se

1 = slight; 2 = moderate; 3 = severe; - = missing; Medication: at time of examination; ld = levodopa; ac = anticholinergic drugs; br = bromocriptine; se = selegiline.

TABLE 1C

Clinical data of the patients suffering from Friedreich's ataxia (FA)

Pat	Age	Sex	Dur	Family history	Ataxia stance	Ataxia arms	Cranial nerves	Tend refl	Babin	Vibr
FA1	21	f	4	+	2	1	n	-	-	0/8
FA2	21	f	6	-	1–2	1–2	n	red	-	5/8
FA3	25	f	8	-	2	1–2	n	-	+	2/8
FA4	23	f	11	+	2–3	2	VIII	-	+	5/8
FA5	29	m	14	-	3	2	n	-	-	2/8
FA6	27	f	15	-	2	1	n	-	-	2/8
FA7	34	m	16	+	2	1–2	n	-	-	2/8
FA8	33	m	17	+	3	2	n	-	-	1/8
FA9	28	m	18	+	3	2	n	-	+	3/8
FA10	35	f	24	+	3	3	VII	-	+	0/8
FA11	34	m	25	+	3	3	n	-	+	0/8

Tend refl = tendon reflexes (legs); Babin = Babinski sign; Vibr = vibration (legs); red = reduced.

TABLE 1D

Clinical data of the patients suffering from purely cerebellar atrophy (CA)

Pat	Age	Sex	Dur	Family history	Ataxia stance	Ataxia arms	Cranial nerves	Tend refl	Babin	Vibr
CA1	36	m	1	+	1	-	n	n	-	n
CA2	76	f	1	-	2	1	VIII	n	-	6/8
CA3	61	f	2	-	1	1	n	n	-	4/8
CA4	56	m	3	-	1	1	n	n	-	n
CA5	54	m	4	-	2	1	n	n	-	n
CA6	59	m	4	-	2	2	n	n	-	3/8
CA7	68	f	5	-	1	-	n	n	-	5/8
CA8	55	m	6	+	2	2	n	n	-	5/8
CA9	64	f	13	+	3	2	n	n	-	4/8
CA10	54	m	14	+	2	1	n	n	-	n
CA11	55	m	17	-	2	2	n	n	-	4/8
CA12	23	m	a	-	1	2	n	n	-	n
CA13	58	m	a	-	1	1	n	n	-	n
CA14	63	f	a	-	1	2–3	n	n	-	6/8

and production of sustained vowels. Two parts of this investigation were considered for acoustic analysis of speech timing.

a) Oral diadochokinesis: Subjects were asked to repeat a given syllable "as fast as possible on a single breath". The following items were considered: /ka/, /pa/, and /ta/. Thus, three repetition trains were obtained from each subject.

b) Articulation test sentences: The test sentences consisted of the nonsense target word "geC1VC2e" embedded in the German carrier phrase "Ich habe ... gehört" ("I have heard ...") (Ziegler and von Cramon 1986; Ackermann and Ziegler 1991). Using the two substitution rules: "C1,C2 = {/p/, /t/, /k/}" and "V = {/i/, /u/, /y/, /a/}", twelve test sentences were derived, e.g. "Ich habe gekake gehört", "Ich habe gekike gehört", "Ich habe gepape gehört" etc. In most German dialects the stop consonants /p/, /t/, /k/, when produced in vowel surrounding, represent fortis, aspirated, and voiceless phonemes. The stressed syllable of the target word, i.e. C1V, bears the main focus of the sentences. These test materials were considered a compromise between naturalness of speech, on the one hand, and the requirements of standardization, on the other. The examiner presented the twelve items orally in randomized order twice. Subjects were asked to repeat each item. Thus, 24 sentence utterances were obtained from each subject.

2.3 *Recording and processing of the speech signal*

Speech examination was done in a sound treated room using a condenser microphone (C-48; Sony, Japan) and either a DAT recorder (PCM 2000; Sony, Japan) or an analogue tape recorder (NAGRA IV-2; Nagra-Kudelski, Switzerland). The mouth-microphone distance amounted to about 20 cm.

Prior to segmentation the recordings were band-pass filtered (50–4000 Hz), digitized at a sampling rate of 10 kHz, and stored on a personal computer (486er IBM-compatible). Acoustic analysis was performed by means of commercially available software (Computerized Speech Lab CSL 4300; Kay Elemetrics Corp., USA) which allowed auditory playback of marked segments. For segmentation the speech signal was displayed on the PC monitor screen using a maximal horizontal resolution of .1 ms and a maximal vertical resolution of 16 bit.

Acoustic measurements were performed at syllables nos. 2 to 6 of each diadochokinesis train and at the central part of the articulation test sentences ("...abe geC1VC2e ge..."). With the support of auditory playback onset and offset of the vocalic segments of the syllable repetitions as well as the target word "geC1VC2e" and the burst onset of the stop consonants C1

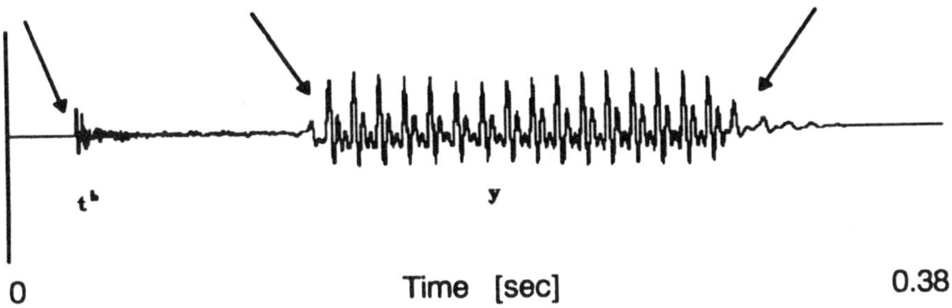

Fig. 1. Oscillogram of the speech signal (C1V of the target word): Arrows indicate (from left to right) stop consonant burst, vowel onset, and vowel end.

and C2 of the target word were visually identified and marked by means of a cursor (Figure 1; for further details see Ackermann and Hertrich 1993; Hertrich and Ackermann 1994).

The following criteria were used for the segmentation of the speech signal: (A) 'Vowel onset' was identified as the beginning of periodic acoustic energy (Lisker and Abramson 1964). Usually, this event coincides with the onset of audible vowel quality. Sometimes patients produced whispered vowels lacking periodicity. Then the earliest point at which the respective vowel quality could be perceived was defined as vowel onset. (B) The steep decline of signal amplitude concomitant with a dampening of higher-frequency components was considered 'vowel end'. (C) Usually, the production of unvoiced, aspirated stop consonants gives rise to a release burst at the speech signal. In instances of missing bursts an 'incomplete closure' was noted. The fortis consonants C1 and C2 (=/k/, /p/, or /t/) sometimes presented with a small burst superimposed on a residual air stream during the closure interval ("spirantization"; Weismer 1984). Spirantization was assumed in the presence of an audible air stream concomitant with noise at the oscillogram during the occlusion period. These stop consonants were considered incomplete as well (Figure 2).

2.4 *Data analysis and statistics*

A former study had shown that FA patients may present with shortened as well as lengthened VOT (Ackermann and Hertrich 1993). In the presence of deviations from the normal range in both directions group means and medians, respectively, can be unremarkable. Therefore, comprehensive individual data of the parameters considered are presented in the Appendices.

In order to assess the specific influence of the various central motor disorders on the acoustic measures considered, analyses of variance were applied to the data. Since, first, durational data, as a rule, are not normally distributed and since, second, the variances may considerably differ between subject groups, Kruskall–Wallis analysis of variance by ranks was applied. This procedure provides a rather conservative criterium of statistical significance.

Tests including multiple comparisons accidentally may give rise to significant effects. Therefore, the Bonferroni–Holm correction for the adjustment of significance levels had to be applied [$alpha^* = alpha/(k - i + 1)$ (k = number of analyses performed, $i = i^{th}$ lowest significance value p)]. Altogether 16 analyses were performed ($n = 4$ with respect to the syllable repetition task; syllabic timing, $n = 3$; intrasyllabic timing, $n = 9$). Assuming an alpha-level of 0.05, thus, the lowest of the 16 p-values had to be $<0.05/16$ (=.0031) in order to be significant, the second-lowest $<0.05/15$ (=.0033), the third-lowest $p < .05/14$ (=.0036) and so on. Significant group effects were further specified by means of post-hoc pairwise comparisons (Kruskall–Wallis analyses with the data from two subject groups each).

To test the effects of 'syllable position within sentence' and 'subject group' simultaneously, a two-factor ANOVA with subsequent multiple range analyses was, in addition, performed with the data on syllable durations. Prior to this analysis the durational parameters were converted onto a logarithmic scale in order to compensate for the skewness of their distribution.

3. SYLLABLE REPETITION TASK

3.1 *Parameters*

a) For each repetition train the number of syllables produced was determined. The individual mean across the three sequences obtained from each subject provided the parameter 'repetitions per train'.

b) The stop consonants considered for analysis are characterized by an occlusion of the oral cavity, performed by the lips in the case of /p/, by the tongue blade in the case of /t/, and by the tongue dorsum in the case of /k/. Subsequent interruption of the oral air stream gives rise to a silent interval followed by a release burst. Incomplete closure was assumed in the presence of an audible air stream concomitant with detectable acoustic energy at the speech oscillogram (Figure 2). Using the first seven stop

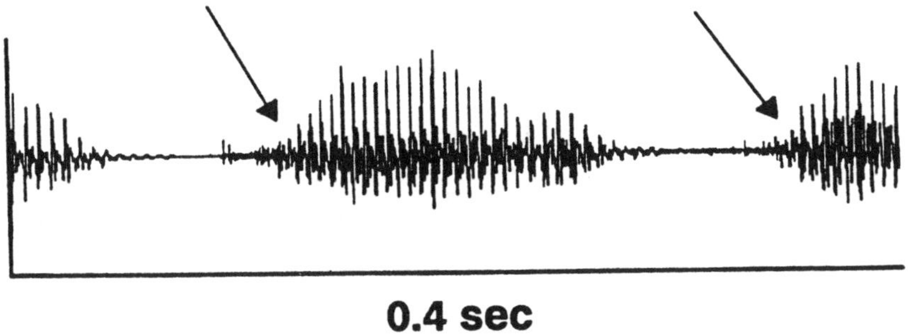

Fig. 2. Exemplification of complete (left) and incomplete (right) stop consonant closure.

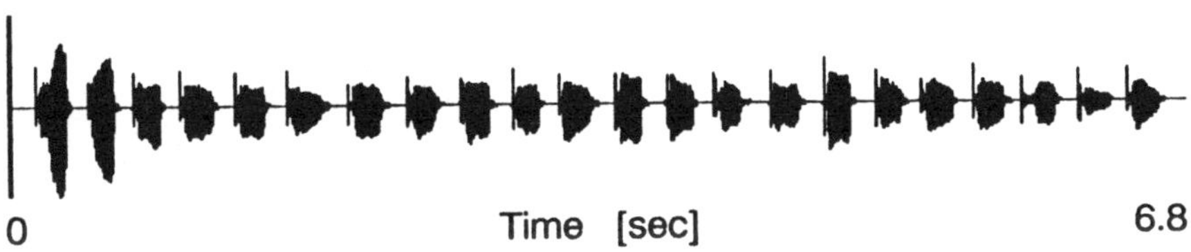

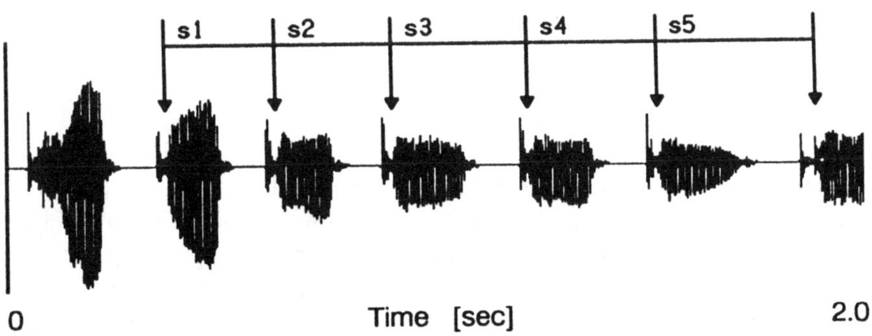

Fig. 3. Upper panel: example of a syllable repetition train. Lower panel: zoomed initial part of this sequence, vowel onset of syllables nos. 2 to 6 being marked.

consonants of the repetition sequences the 'percentage of incomplete stop consonant occlusions' across the three syllable repetition trains of each subject was determined.

c) Syllable duration was defined as the interval between two successive vowel onsets. The first syllable was excluded from analysis since the control subjects often produced it with longer duration than the succeeding ones. A number of patients presented with rather short repetition trains. In order to obtain comparable data sets, only five syllables were considered for analysis of syllable lengths. With respect to each repetition train, thus, the median duration of syllables no. 2 to 6 was determined (Figure 3). The parameter 'median syllable duration' was defined as the individual mean value across the three medians obtained from each subject. Vowel end of syllable no. 6 coincides with the onset of syllable no. 7. Trials comprising less than seven syllables were handled as missing values. Due to eventual intrinsic effects of the consonants /k/, /p/, and /t/ upon syllable length, missing values might result in a systematic bias of the computed means. Median syllable durations, therefore, were normalized for consonant effects by adding the respective residuals — as obtained by an analysis of variance — to the overall mean value.

d) The standard deviation of syllable durations may depend on the respective absolute mean value (Crystal and House 1988). Therefore, variation coefficients (= standard deviation divided by mean) were used as measures of durational variability across syllables nos. 2 to 6 of each repetition train. The computed mean of the three (/k/, /p/, /t/) variation coefficients obtained from each subject provided the parameter 'intra-train variation coefficient of syllable durations'.

3.2 *Results*

Inspection of the individual data shows that in all patients — apart from a few subjects in a mild state of their disease — the value of at least one of the four parameters was outside the normal range (Appendix 1). Using Kruskall–Wallis analysis of variance significant group effects were found with respect to all four parameters. Two patient groups showed rather specific profiles as revealed by post-hoc pairwise comparisons (Table 2; see Figures 4 and 5): (A) The PD subjects presented with normal train lengths, unimpaired rate in terms of median syllable duration and a high percentage of incomplete closures concomitant with increased variability of syllable durations. (B) In contrast, the FA group had rather stable syllable lengths and complete occlusions, but showed the shortest trains and the lowest syllable repetition rates. As a rule, HD patients had the highest intra-train variability of syllable durations.

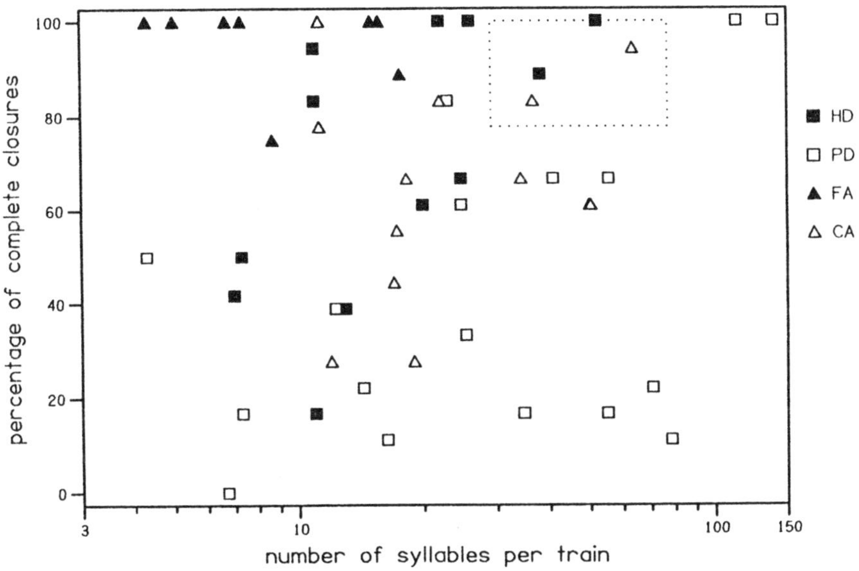

Fig. 4. Length of the syllable repetition trains (number of syllables per sequence) versus percentage of complete stop consonant closures; each point represents one subject (dotted box = normal range).

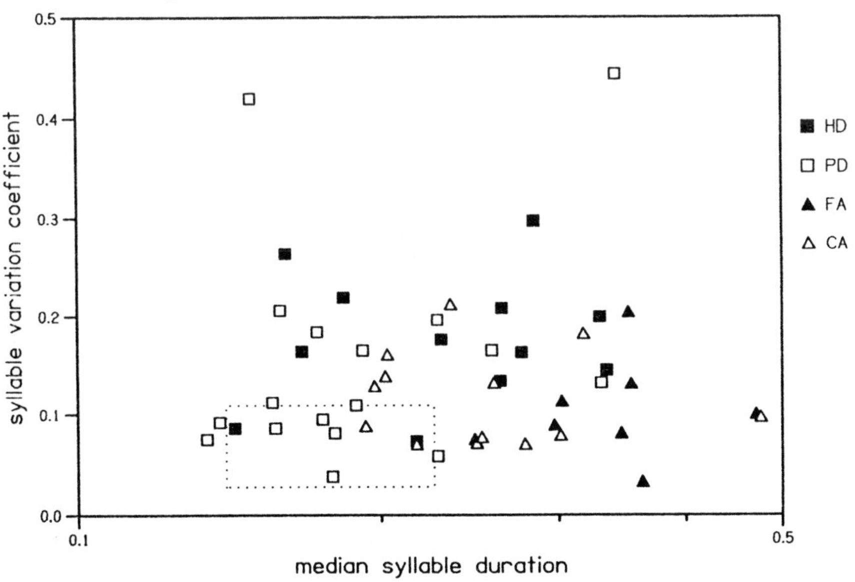

Fig. 5. Median syllable duration versus variation coefficient of syllable duration (syllable repetition task); each point represents one subject (dotted box = normal range).

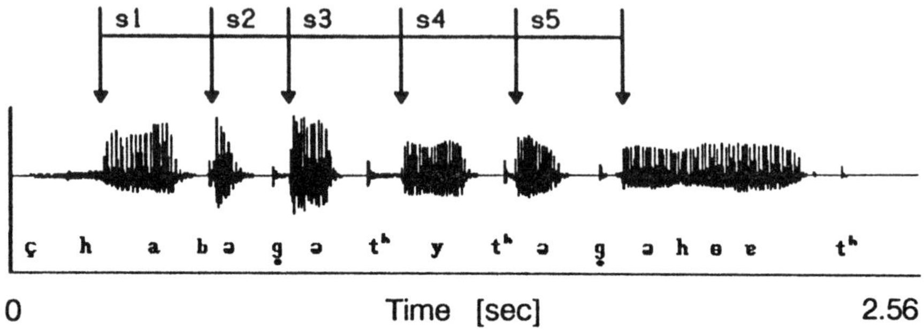

Fig. 6. Syllable durations (s1 to s5) of an articulation test sentence.

4. SENTENCE UTTERANCES

4.1 *Syllabic timing*

4.1.1 *Parameters*

a) Syllable durations were determined from vowel onset to vowel onset each. The following five syllabic segments were considered for analysis: /ab/, /eg/, /eC1/, /VC2/, /eg/ (s1 to s5 in Figure 6). Utterances including sound elisions or syllable repetitions were handled as missing values. The numbers of test sentences considered for analysis in each subject are shown in Appendix 2. The most severely impaired HD patient (HD14) produced truncated test sentences concomitant with severely reduced consonants, giving rise to perceived diphthongs instead of vowel-stop-vowel sequences. Therefore acoustic analysis did not seem feasible in this subject. Due to technical problems PD16 performed the articulation test once only.

Because of variable performance in all speaker groups, e.g. omission of /i/ in "Ich" or missing stop consonant release burst of utterance-final /t/, the initial and the final part of the sentences were excluded from analysis. Admittedly, the beginning of a syllable does not necessarily coincide with vowel onset. Two arguments, however, favor the approach chosen in the present study. First, vowel onset represented the most reliably identifiable event at the acoustic speech signal of the articulation test utterances. Second, it seems to be the relevant landmark of a syllable regarding the coordination of phonetic segments with global sentence structure (Browman and Goldstein 1988). The medians of the five syllables s1 to s5 each

TABLE 2

Parameters of oral diadochokinesis (rapid syllable repetitions): Kruskall–Wallis analysis of variance by ranks

A) Number of syllables per train

Group	N	Average rank	Post-hoc pairwise comparison HD	PD	FA	CA
NO	15	50.9	**	ns	**	**
HD	14	22.8		*	ns	ns
PD	17	42.0			**	ns
FA	9	12.2				**
CA	13	33.7				

Test statistic = 29.10; significance level = .000007

B) Percentage of incomplete stop consonant closures

Group	N	Average rank	Post-hoc pairwise comparison HD	PD	FA	CA
NO	15	17.5	*	**	ns	**
HD	14	34.7		*	ns	ns
PD	17	50.1			**	*
FA	9	21.2				*
CA	13	39.9				

Test statistic = 27.85; significance level = 0.00001

C) Median syllable duration

Group	N	Average rank	Post-hoc pairwise comparison HD	PD	FA	CA
NO	15	20.7	*	ns	**	**
HD	12	37.2		*	**	ns
PD	17	21.0			**	**
FA	8	57.6				*
CA	13	43.5				

Test statistic = 30.99; significance level = .000003

D) Variation coefficient of syllable duration

Group	N	Average rank	Post-hoc pairwise comparison HD	PD	FA	CA
NO	15	16.4	**	**	ns	**
HD	12	49.1		ns	*	*
PD	17	37.2			ns	ns
FA	8	30.3				ns
CA	13	33.2				

Test statistic = 21.25; significance level = .00028

The matrices at the right side of the table show the results of post-hoc pairwise comparisons. N = sample size; ns = not significant; * = $p < 0.5$; ** = $p < .01$.

across 24 sentence repetitions as well as their sum, i.e. utterance duration, were considered measures of speech tempo.

b) Two parameters of intra-individual variability of speech tempo across sentence repetitions were introduced: the 'variation coefficient of utterance duration', i.e. the standard deviation of utterance duration divided by the respective mean value, and the 'mean syllable variation coefficient', i.e. the mean value of the variation coefficients of syllables s1 to s5 obtained from each subject. Admittedly, syllables nos. 3 and 4, i.e. eC1 and VC2, vary with respect to their phonological content across the 24 utterances obtained from each individual whereas syllables nos. 1, 2 and 5 are invariant in these regards. However, no significant differences between phonologically 'invariant' and 'variant' syllables were found (t-test, $p > .05$). Moreover, the intrinsic consonant and vowel effects upon syllable duration proved to be rather small as compared to inter-individual variation or as compared to the influence of the factor 'diagnostic group'.

c) Principally, sentence utterances show a tendency towards syllable isochrony. However, due to prosodic and phonological demands such as word stress or contrast between German long and short vowels syllables actually do not present with equal lengths. Increased syllabic isochrony ('scanning speech') has been reported in cerebellar and bulbocerebellar dysfunctions (Kent et al. 1979; Kent and Rosenbek 1982). As a measure of intra-utterance variability of syllable durations the 'durational scanning index' was introduced (Ackermann and Hertrich 1993):

$$\text{Scanning index} = \frac{s1 * s2 * s3 * s4 * s5}{[(s1 + s2 + s3 + s4 + s5)/5]^5}.$$

Ackermann and Hertrich (1993) applied the scanning index formula to each single utterance. The present study, in contrast, used the median durations of the syllables s1 to s5 across 24 repetitions. This procedure counter-balances to some extent the effect of increased inter-utterance variability.

4.1.2 *Results*

Measures of speech tempo. Appendix 2 presents the individual median values of syllable and utterance durations. A considerable number of HD, FA, and CA patients had utterance lengths exceeding the normal range. In contrast, two PD subjects only showed slowed speech tempo. Two further individuals presented with even increased speaking rate. Syllable s5, comprising the post-accent vowel 'e' of the target word and the following inter-word pause was most often lengthened in the various patient groups.

A two-factor analysis of variance was performed in order to examine the effects of 'syllable' (s1, s2, s3, s4, s5) and 'diagnostic group' (NO, HD, PD, FA, CA) on syllable lengths. Both factors showed significant effects upon median syllable duration (Table 3). Subsequent multiple range analyses revealed that syllables nos. 3 and 4, which comprise the domain of the sentence focus C1V, were significantly longer than syllables nos. 1, 2 and 5. As concerns factor 'diagnostic group' the CA and FA patients had significantly lengthened syllables as compared to all other subject groups. The HD patients had significantly longer durations than the controls and the PD subjects. In contrast, the PD patients did not present with significantly increased median syllable lengths as compared with the normal speakers.

In accordance with these results Kruskall–Wallis analysis of variance by ranks revealed a significant effect of the factor 'group' on median utterance duration. Post-hoc pairwise comparisons showed significantly increased utterance durations of the CA, FA and HD patients in comparison with the controls and the PD group (Table 4A).

Inter-utterance variability. The same procedure as performed with median syllable durations was applied to their variation coefficients. Both factors, i.e. 'syllable' and 'diagnostic group', showed significant effects (Table 5). Multiple range analyses revealed that syllable s5 had the highest variation coefficient whereas syllable s1 was the most invariant one. Despite their variable phonetic content syllables s3 and s4 did not present with increased variation coefficients. The effects of the various phonemes (/k/, /p/, /t/ and /a/, /i/, /u/, /y/, respectively) on the syllable variation coefficients, thus, can be neglected. All four patient groups had significantly increased syllable variation coefficients as compared to the control group. Furthermore, the two-factor interaction achieved significance, indicating that the variation coefficients of the five syllables had different distributions within the various groups. This effect predominantly was due to the overproportionally increased variability of syllable s3 ("eC1") in the two ataxic groups.

Kruskall–Wallis analysis of variance by ranks performed with the mean syllable variation coefficients revealed a significant effect of factor 'group' as well. Post-hoc pairwise comparisons showed increased values of the four patient groups as compared to the controls (Table 4B). However, no significant differences between them emerged. The variation coefficient of utterance duration showed a highly significant correlation with the mean syllable variation coefficient (Spearman rank correlation coefficient $r = .96$; $p<.0001$) and, thus, was not further examined.

Intra-utterance variability. Kruskall–Wallis analysis of variance by

TABLE 3

Analysis of variance of syllable durations (logarithms)

Source of variation	Sum of Squares	d.f.	Mean square	F-ratio	Sig. level
MAIN EFFECTS	16.454484	8	2.0568104	21.340	.0000
syllable no.	5.180214	4	1.2950536	13.436	.0000
subject group	11.274269	4	2.8185673	29.243	.0000
2-FACTOR INTERACTIONS	1.2811478	16	0.0800717	0.831	.6503
RESIDUAL	31.325146	325	0.0963851		
TOTAL (corr.)	49.060777	349			

Multiple range analysis based on 95% confidence intervals

Syllable effect:

Syllable number	Count	Average	Homogeneous groups
s2	70	-1.6898571	*
s1	70	-1.6014979	*
s5	70	-1.6006528	*
s4	70	-1.4338874	*
s3	70	-1.3586914	*

Group effect:

Subject group	Count	Average	Homogeneous groups
NO	75	-1.7376727	*
PD	85	-1.6946054	*
HD	65	-1.5193713	*
CA	70	-1.3487659	*
FA	55	-1.2796619	*

TABLE 4

Parameters of syllabic timing: Kruskall–Wallis analysis of variance by ranks

A) Median utterance duration

Group	N	Average rank	Post-hoc pairwise comparison HD	PD	FA	CA
NO	15	22.1	*	ns	**	**
HD	13	39.4		*	ns	ns
PD	17	23.7			**	**
FA	11	51.2				ns
CA	14	48.0				

Test statistic = 24.51; significance level = .00006

B) Mean syllable variation coefficient

Group	N	Average rank	Post-hoc pairwise comparison HD	PD	FA	CA
NO	15	11.1	**	**	**	**
HD	13	46.0		ns	ns	ns
PD	17	42.5			ns	ns
FA	11	37.3				ns
CA	14	41.7				

Test statistic = 28.47; significance level = .00001

C) Durational scanning index

Group	N	Average rank	Post-hoc pairwise comparison HD	PD	FA	CA
NO	15	24.6	ns	ns	*	**
HD	13	32.9		ns	ns	ns
PD	17	31.0			ns	*
FA	11	44.8				ns
CA	14	47.6				

Test statistic = 12.61; significance level = 0.013

D) Durational scanning index: FA and CA versus all other subjects

Group	N	Average rank
NO, HD, PD	45	29.4
FA, CA	25	46.4

Test statistic = 11.16; significance level = .00084

The matrices at the right side of the table show the results of post-hoc pairwise comparisons. N = sample size; ns = not significant; * = $p < 0.5$; ** = $p < .01$.

TABLE 5

Analysis of variance of syllable variation coefficients (logarithms)

Source of variation	Sum of Squares	d.f.	Mean square	F-ratio	Sig. level
MAIN EFFECTS	29.516865	8	3.6896081	18.214	.0000
syllable no.	10.159191	4	2.5397978	12.538	.0000
subject group	19.357673	4	4.8394183	23.89	.0000
2-FACTOR INTERACTIONS	8.7566902	16	0.5472931	2.702	.0005
RESIDUAL	65.834361	325	0.2025673		
TOTAL (corr.)	104.10792	349			

Multiple range analysis based on 95% confidence intervals

Syllable effect:				Group effect:			
Syllable number	Count	Average	Homogeneous groups	Subject group	Count	Average	Homogeneous groups
s1	70	-2.0875491	*	NO	75	-2.2515600	*
s4	70	-1.8920111	**	FA	55	-1.7323389	*
s3	70	-1.7662342	**	CA	70	-1.7285764	*
s2	70	-1.6987810	**	PD	85	-1.6717991	*
s5	70	-1.5929330	*	HD	65	-1.6211797	*

ranks detected a significant, but rather small influence of the factor 'subject group' on the scanning index (Table 4C). Assuming that scanning speech primarily represents a sign of ataxic dysarthria, the data from the FA and CA groups were pooled and tested against all other subjects. Under these conditions the ataxic patients presented with significant durational scanning (Table 4D).

4.2 *Intra-syllabic segments*

4.2.1 *Parameters*

(a) The individual median voice-onset-time ('median VOT') of the stop consonant C1 was computed across all three categories considered, i.e. /p/, /t/, and /k/. Incomplete closures (controls 6.7%, HD 13.2%, PD 38.5%, FA 5.7%, CA 13.1% of all C1 productions) were excluded from analysis (for definition of incomplete closure see Section 2.3). In order to correct the VOT data for intrinsic consonant effects, an analysis of variance was performed and the residuals added to the overall mean (Hertrich and Ackermann 1994).

(b) The standard deviation of VOT showed a significant correlation with utterance length ($r = .76$; $p < .0001$). Therefore, variation coefficients, i.e. standard deviation divided by the respective mean, were used as measures of intra-individual variability of VOT across repetitions. For each subject the variation coefficient of /k/, /p/, and /t/, respectively, was determined and the mean across these three values computed ('VOT variation coefficient').

(c) To evaluate the relative strength of intrinsic consonant effects, the 'VOT difference between /k/ and /p/' was determined. As a rule, these two categories are characterized by the longest and the shortest intrinsic duration, respectively.

(d) The median durations of four intra-syllabic segments (V = duration of the target vowel V, schwa = length of the post-target schwa, C1, C2 = duration of the inter-vocalic segment of the stop consonants C1 and C2, respectively) across the 24 test sentence productions were computed (Figure 7). The latter two intervals comprise stop consonant occlusion plus succeeding VOT. In some instances the vowel V was whispered, i.e. produced completely voiceless without any audible and/or visible periodicity at the acoustic signal (controls 0%, HD 0.7%, PD 4%, FA 2.3%, CA 2.1% of all V productions). These cases were handled as missing values in this context.

(e) The ratio (C1+V)/(C1+V+C2+schwa) was considered a measure of

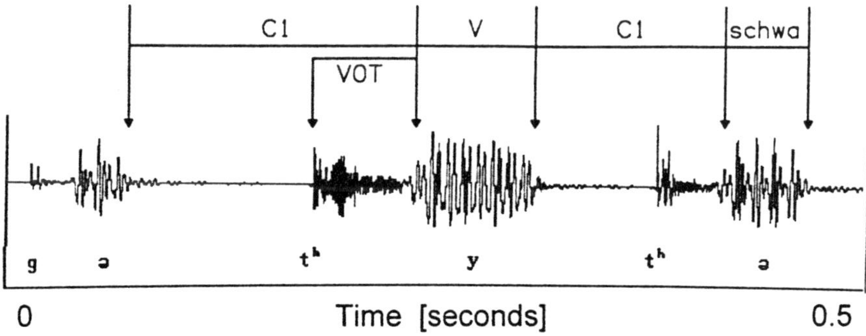

Fig. 7. Example of intra-syllabic segments in the zoomed target word "getyte" uttered by a male control subject.

durational stress contrast within the target word of the articulation test sentences ('stress').

(f) The measure 'vocpart', i.e. the ratio (V+schwa)/(C1+V+C2+schwa), was introduced to assess the proportion of vocalic segments within the target word.

4.2.2 *Results*

Using the independent factor variable 'group' (controls, HD, PD, FA, CA), a Kruskall–Wallis analysis of variance by ranks was performed with each of these measures of intrasyllabic timing. Since all comparisons achieved significance, pairwise post-hoc tests were added in order to further specify the group effects.

Both the HD and the PD patients had significantly reduced median VOT (Table 6A). In contrast, the rank test did not reveal significant differences with respect to VOT between either ataxic group and the controls. Visual inspection of the individual data showed enlarged inter-subject variability of VOT in the ataxic subjects with a considerable number of either decreased or increased values as compared to the controls (Appendix 3).

Among the various VOT parameters variability across sentence repetitions revealed to be the most strongly affected measure in all patient groups (Table 6B). Nearly all patients had values exceeding the normal range (Appendix 3). VOT variability increased in the following order: PD, CA, HD, FA.

The ataxic groups had an increased difference between /k/- and /p/-

TABLE 6

VOT data: Kruskall–Wallis analysis of variance by ranks

A) Median VOT (A.vot1med)

Group	N	Average rank	Post-hoc pairwise comparison HD	PD	FA	CA
NO	15	41.3	**	*	ns	ns
HD	13	21.9		ns	ns	**
PD	15	27.6			ns	*
FA	11	35.2				ns
CA	14	45.7				

Test statistic = 13.40; significance level = .0095

B) Variation coefficient of VOT

Group	N	Average rank	Post-hoc pairwise comparison HD	PD	FA	CA
NO	15	11.2	**	**	**	**
HD	13	48.2		**	ns	ns
PD	15	28.1			**	*
FA	11	52.0				*
CA	14	39.9				

Test statistic = 38.33; significance level < .00001

C) VOT difference between /k/ and /p/

Group	N	Average rank	Post-hoc pairwise comparison HD	PD	FA	CA
NO	15	24.7	ns	ns	*	*
HD	13	24.9		ns	ns	*
PD	15	24.9			*	*
FA	11	40.6				ns
CA	14	40.4				

Test statistic = 11.33; significance level = .023

VOT as compared both to the controls and the patients with basal ganglia dysfunction (Table 6C). Eight of the 24 ataxic subjects presented with median /k/-VOT exceeding the normal range whereas a single individual only had a decreased value. In contrast, /p/-VOT was abnormally prolonged in four and shortened in seven cases (Appendix 3).

All patient groups had a significantly lengthened final schwa of the target word as compared to the controls (Table 7A). In spite of their normal speech tempo, this holds true even for the PD subjects. As concerns V, the controls presented with the shortest median durations. However, only the

TABLE 7

Intra-syllabic segments V, schwa, C1, and C2: Kruskall–Wallis analysis of variance by ranks

A) Median duration of schwa

Group	N	Average rank	Post-hoc pairwise comparison HD	PD	FA	CA
NO	15	15.6	**	**	*	**
HD	13	42.0		ns	ns	ns
PD	17	32.0			ns	*
FA	11	45.2				ns
CA	14	47.3				

Test statistic = 23.36; significance level = .0001

B) Median duration of target vowel V

Group	N	Average rank	Post-hoc pairwise comparison HD	PD	FA	CA
NO	15	22.5	ns	ns	**	**
HD	13	37.1		ns	ns	ns
PD	17	30.2			ns	*
FA	11	45.6				ns
CA	14	46.4				

Test statistic = 14.04; significance level = .0072

C) Median duration of C1

Group	N	Average rank	Post-hoc pairwise comparison HD	PD	FA	CA
NO	15	31.0	ns	ns	**	*
HD	13	36.1		*	ns	ns
PD	17	20.1			**	**
FA	11	50.8				ns
CA	14	45.4				

Test statistic = 18.97; significance level = .0008

D) Median duration of C2

Group	N	Average rank	Post hoc pairwise comparison HD	PD	FA	CA
NO	15	30.0	ns	ns	**	*
HD	13	34.9		ns	*	ns
PD	17	21.7			**	**
FA	11	51.2				ns
CA	14	46.4				

Test statistic = 19.48; significance level = .0006

TABLE 7

Continued

E) Vocalic part

Group	N	Average rank	Post-hoc pairwise comparison HD	PD	FA	CA
NO	15	22.6	*	**	ns	ns
HD	13	42.1		ns	ns	ns
PD	17	45.6			ns	ns
FA	11	34.4				ns
CA	14	31.8				

Test statistic = 12.11; significance level = 0.016

F) Stressed part

Group	N	Average rank	Post-hoc pairwise comparison HD	PD	FA	CA
NO	15	45.6	**	ns	*	ns
HD	13	24.0		ns	ns	ns
PD	17	37.9			ns	ns
FA	11	28.4				ns
CA	14	38.0				

Test statistic = 9.65; significance level = .047

differences between controls, on the one hand, and the CA and FA groups, on the other, achieved significance (Table 7B; Appendix 4).

The PD subjects presented with the shortest duration of both consonant segments among all groups including the controls. In contrast, the ataxic patients had significantly lengthened C1 and C2 intervals. Moreovoer, the same rank order across the various groups emerged both for C1 and C2 (Tables 7C and 7D; Appendix 4).

The PD as well as the HD subjects presented with a significantly increased 'vocalic part' of the segments considered (Table 7E). Furthermore, all patient groups had a shorter relative duration of the accent syllable C1V than the controls. This effect achieved significance in the HD and the FA groups (Table 7F).

5. DISCUSSION

The present study performed measurements of syllabic and intra-syllabic segments at the acoustic speech signal in patients with Parkinson's (PD) and Huntington's disease (HD), Friedreich's ataxia (FA) and purely cerebellar atrophy (CA). Table 8 summarizes the profiles of performance during

TABLE 8
Profiles of performance during syllable repetition and sentence repetition tasks

Parameter	HD	PD	FA	CA
Diadochokinesis tasks:				
short syllable trains	++	o	++	+
incomplete stop closures	+	++	o	+
slow syllable rate	+	o	++	++
durational irregularity	+	o	++	++
Sentence utterances:				
slowed speech rate	+	o	++	++
irregular syllable duration	++	++	++	++
scanning index	o	o	+	+
increased vocalic part	+	++	o	o
reduced durational stress marking	++	o	+	o
shortened VOT	++	+	o	o
variable VOT	++	+	++	++
increased consonant effect upon VOT	o	o	+	+

oral diadochokinesis and repetition of articulation test sentences as derived from both the inspection of the individual data and the significance tests performed. Three major findings emerged: First, all four neurological dysarthrias considered gave rise to increased variability of VOT and syllable lengths across sentence repetitions. Second, characteristic differences between PD subjects and the other patient groups emerged with respect to measures of speech tempo and the length of produced syllable repetition trains. Third, the ataxic patients, on the one hand, and the basal ganglia dysfunctions, on the other, showed different constellations in most of the remaining parameters of syllabic and intrasyllabic timing, i.e. scanning speech, VOT and the proportion of vocalic segments.

In consideration of the sparse kinematic and electromyographic data on dysarthric performance available so far, interpretation of the obtained acoustic findings in terms of pathophysiology must remain premature and speculative. Therefore only a few comments will be added. Increased variability of syllabic and intrasyllabic segments across sentence repetitions seems to be a rather universal and, thus, unspecific feature of neurological dysarthrias. These findings might simply reflect a decreased 'signal-to-noise ratio' due to the fact that the speech apparatus is strained beyond its optimal working range and, thus, more susceptible to random perturbations. It cannot be excluded, however, that some specific pathomechanisms contribute to the observed durational irregularities. There is some experimental evidence that the cerebellum represents an internal clock required, among others, for the timing of motor events (Keele and Ivry 1991). Thus, disordered clock functions might account for the increased variability of VOT and syllable lengths across sentence repetitions in ataxic patients. The main source of irregularity in the Huntingtonian group, in contrast, could be either the chorea itself or an increased effort to suppress choreatic activity.

Speech rate differentiated between PD subjects, on the one hand, and the HD, CA and FA groups, on the other. In contrast to PD the latter disorders gave rise to slowed speech tempo in terms of prolonged utterance durations as well as lengthened syllables and vowels. Both cerebellar and basal ganglia dysfunctions may result in slowed motor execution (bradykinesia) and/or delayed transitions between successive movements (Benecke et al. 1987; Thompson et al. 1988). Assuming that these disorders affect the speech motor system as well, reduced speech rate, thus, can be expected in cerebellar and basal ganglia disorders. Nevertheless, PD patients, as a group, had normal speech rate both during diadochokinesis tasks and sentence repetitions. Kinematic and acoustic measurements provided some evidence that PD subjects produce "undershooting" articulatory gestures (Caligiuri 1987; Connor et al. 1989a; Forrest et al. 1989; Ackermann and Ziegler 1991). The remarkably high amount of incomplete closures in PD subjects in the present study is in accordance with these data. So far it is unclear, however, why patients with Parkinson's disease reduce movement amplitude whereas patients with other central motor disorders perform with slow speech rate.

It is noteworthy that two PD patients consistently presented with reduced syllable rate both during syllable repetition and sentence utterances. A rather small subgroup of PD patients with reduced speech tempo has been observed by Ludlow and coworkers (1987) as well. So far it is unsettled whether these patients represent a distinct subgroup of PD patients.

VOT, i.e. the time-interval between stop consonant burst and onset of the succeeding vowel (Lisker and Abramson 1964), has been considered a parameter of the temporal coordination of laryngeal and orofacial gestures during speech (Keller 1990). In accordance with this suggestion there seems to be no correlation between VOT and the duration of other speech segments as e.g. vowel length (Port and Rotunno 1979). This parameter provides phonological information with respect to stop consonant categories as e.g. the /d/-/t/ distinction. As a group both the HD and PD patients presented with shortened VOT. These findings are in accordance with the observation of an increased overlap of voiced and unvoiced stop consonant productions with respect to VOT in PD (Lieberman et al. 1992). Investigations of upper limb motor activities revealed an impaired ability to rapidly develop isometric forces in HD patients (Hefter et al. 1987). In consideration of these findings shortened VOT in HD might reflect reduced muscular tension during stop consonant production giving rise to faster release of the occlusion and, thus, earlier vowel onset. Lieberman and coworkers considered the "VOT timing errors" in PD as deficits of prefrontal language processing due to basal ganglia dysfunction. As an alternative, shortened VOT in PD might reflect hypometria of the de-voicing gestures during the production of voiceless stops. At some variance with the results of the present study Connor et al. (1989b) reported unimpaired VOT in Parkinsonian patients.

The ataxic subjects showed both shortened and lengthened absolute VOT. Kent and coworkers (1979) observed deviations of VOT in either direction from the normal range as well. Conceivably, systematic shifts of the assumed cerebellar clock (see above) account for these deviations. Alternatively, lengthened absolute VOT could reflect delayed initiation of laryngeal oscillations whereas VOT shortening might be due — as suggested in HD — to reduced isometric force generation (see above).

REFERENCES

ACKERMANN, H., ZIEGLER, W. (1991) Articulatory deficits in Parkinsonian dysarthria: An acoustic analysis. *Journal of Neurology, Neurosurgery and Psychiatry*, 54, 1093-1098.

ACKERMANN, H., HERTRICH, I. (1993) Dysarthria in Friedreich's ataxia: Timing of speech segments. *Clinical Linguistics and Phonetics*, 7, 75-91.

ACKERMANN, H., HERTRICH, I. (1994) Speech rate and rhythm in cerebellar dysarthria: An acoustic analysis of syllabic timing. *Folia Phoniatrica et Logopaedica*, 46, 70-78.

BENECKE, R., ROTHWELL, J., DICK, J., DAY, B., MARSDEN, C. (1987) Disturbance of sequential movement in patients with Parkinson's disease. *Brain*, 110, 361-379.

BROWMAN, C., GOLDSTEIN, L. (1988) Some notes on syllable structure in articulatory phonology. *Phonetica*, 45, 140-155.

CALIGIURI, M. (1989) The influence of speaking rate on articulatory hypokinesia in parkinsonian dysarthria. *Brain and Language*, 36, 493-502.

CONNOR, N., ABBS, J., COLE, K., GRACCO, V. (1989a) Parkinsonian deficits in serial multiarticulate movements for speech. *Brain*, 112, 997-1009.

CONNOR, N., LUDLOW, C., SCHULTZ, G. (1989b) Stop consonant production in isolated and repeated syllables in Parkinson's disease. *Neuropsychologia*, 27, 829-838.

CRYSTAL, T., HOUSE, A. (1988) A note on the variability of timing control. *Journal of Speech and Hearing Research*, 31, 497-502.

DARLEY, F., ARONSON, A., BROWN, J. (1969a) Differential diagnostic patterns of dysarthria. *Journal of Speech and Hearing Research*, 12, 246-269.

DARLEY, F., ARONSON, A., BROWN, J. (1969b) Clusters of deviant speech dimensions in the dysarthrias. *Journal of Speech and Hearing Research*, 12, 462-496.

DARLEY, F., ARONSON, A., BROWN, J. (1975) *Motor Speech Disorders*. Philadelphia, PA/USA: Saunders.

FORREST, K., WEISMER, G., TURNER, G. (1989) Kinematic, acoustic, and perceptual analyses of connected speech produced by Parkinsonian and normal geriatric adults. *Journal of the Acoustic Society of America*, 85, 2608-2622.

GENTIL, M. (1990a) Acoustic characteristics of speech in Friedreich's disease. *Folia Phoniatrica*, 42, 125-134.

GENTIL, M. (1990b) Dysarthria in Friedreich disease. *Brain and Language*, 38, 438-448.

HARDING, A. (1984) *The Hereditary Ataxias and Related Disorders*. Edinburgh/UK: Churchill Livingstone.

HEFTER, H., HÖMBERG, V., FREUND, H. (1987) Impairment of fastest movement in Huntington's disease. *Brain*, 110, 585-612.

HERTRICH, I., ACKERMANN, H. (1994) Acoustic analysis of speech timing in Huntington's disease. *Brain and Language*, 47, 182-196.

HIROSE, H., KIRITANI, S., USHIJIMA, T., YOSHIOKA, H., SAWASHIMA, M. (1981) Patterns of dysarthric movements in patients with parkinsonism. *Folia Phoniatrica*, 33, 204-215.

KEELE, S., IVRY, R. (1991) Does the cerebellum provide a common computation for diverse tasks? A timing hypothesis. *Annals of the New York Academy of Sciences*, 608, 179-211.

KELLER, E. (1990) Speech motor timing. In Hardcastle, W., Marchal, A. (Eds), *Speech Production and Speech Modelling*. Dordrecht: Kluwer, 343-364.

KENT, R., NETSELL, R. (1975) A case study of an ataxic dysarthric: Cineradiographic and spectrographic observations. *Journal of Speech and Hearing Disorders*, 40, 115-134.

KENT, R., ROSENBEK, J. (1982) Prosodic disturbance and neurologic lesion. *Brain and Language*, 15, 259-291.

KENT, R., NETSELL, R., ABBS, J. (1979) Acoustic characteristics of dysarthria associated with cerebellar disease. *Journal of Speech and Hearing Research*, 22, 627-648.

KREUL, E. (1972) Neuromuscular control examination (NMC) for Parkinsonism: Vowel prolongations and diadochokinetic and reading rates. *Journal of Speech and Hearing Research*, 15, 72-83.

LIEBERMAN, P., KAKO, E., FRIEDMAN, J., TAJCHMAN, G., FELDMAN, L., JIMINEZ, E. (1992) Speech production, syntax comprehension, and cognitive deficits in Parkinson's disease. *Brain and Language*, 43, 169-189.

LISKER, L., ABRAMSON, A. (1964) A cross-language study of voicing in initial stops: Acoustical measurements. *Word*, 20, 384-422.

LUDLOW, C., BASSICH, C. (1984) Relationships between perceptual ratings and acoustic measures of hypokinetic speech. In McNEIL, M., ROSENBEK, J., ARONSON, A. (Eds), *The Dysarthrias: Physiology, Acoustics, Perception, Management*. San Diego, CA/USA: College-Hill, 163-195.

LUDLOW, C., CONNOR, N., BASSICH, C. (1987) Speech timing in Parkinson's and Huntington's disease. *Brain and Language*, 32, 195-214.

PORT, R., ROTUNNO, R. (1979) Relation between voice onset time and vowel duration. *Journal of the Acoustical Society of America*, 66, 654-662.

PORTNOY, R., ARONSON, A. (1982) Diadochokinetic syllable rate and regularity in normal and in spastic and ataxic dysarthric subjects. *Journal of Speech and Hearing Disorders*, 47, 324-327.

SHOULSON, I., FAHN, S. (1979) Huntington disease: Clinical care and evaluation. *Neurology*, 29, 1-3.

THOMPSON, P., BERARDELLI, A., ROTHWELL, J., DAY, B., DICK, J., BENECKE, R., MARSDEN, C. (1988) The coexistence of bradykinesia and chorea in Huntington's disease and its implications for theories of basal ganglia control of movement. *Brain*, 111, 223-244.

VOLKMANN, J., HEFTER, H., LANGE, H., FREUND, H. (1992) Impairment of temporal organization of speech in basal ganglia diseases. *Brain and Language*, 43, 386-399.

WEISMER, G. (1984) Articulatory characteristics of Parkinsonian dysarthria: Segmental and phrase-level timing, spirantization, and glottal-supraglottal coordination. In Mc-NEIL, M., ROSENBEK, J., ARONSON, A. (Eds), *The Dysarthrias: Physiology, Acoustics, Perception, Management*. San Diego, CA/USA: College-Hill, 101-130.

ZIEGLER, W., VON CRAMON, D. (1986) Spastic dysarthria after acquired brain injury: An acoustic study. *British Journal of Disorders of Communication*, 21, 173-187.

ZYSKI, B., WEISIGER, B. (1987) Identification of dysarthria types based on perceptual analysis. *Journal of Communication Disorders*, 20, 367-378.

APPENDIX 1

Individual data obtained from rapid syllable repetition tasks

Sub	Inc	Nmax	Sylmed	Sylvar
NO1	**0**	34	0.177	0.061
NO2	**19**	55	0.178	0.071
NO3	0	35	0.186	0.054
NO4	0	**29**	0.200	0.070
NO5	19	39	0.167	0.044
NO6	14	44	**0.140**	0.078
NO7	0	35	0.190	0.097
NO8	0	77	0.199	**0.110**
NO9	10	50	0.153	0.028
NO10	19	51	0.179	0.107
NO11	0	53	0.208	0.077
NO12	14	50	**0.227**	**0.040**
NO13	14	52	0.151	0.081
NO14	14	32	0.173	0.090
NO15	0	**149**	0.164	0.055
HD1	0	52	0.161	0.264*
HD2	19	38	0.143	0.087
HD3	33*	20#	0.329*	0.199*
HD4	76*	13#	0.167	0.165*
HD5	0	22#	0.284*	0.297*
HD6	26*	11#	0.335*	0.145*
HD7	24*	24#	0.263*	0.134*
HD8	0	26#	0.218	0.074
HD9	58*	7#	0.276*	0.163*
HD10		1#		
HD11	74*	7#	0.264*	0.209*
HD12	67*	5#		
HD13	5	11#	0.230*	0.177*
HD14	90*	11#	0.184	0.220*

APPENDIX 1

Continued.

Sub	Inc	Nmax	Sylmed	Sylvar
PD1	33*	41	0.180	0.081
PD2	14	139	0.135#	0.076
PD3	19	23#	0.229*	0.058
PD4	95*	78	0.150	0.421*
PD5	71*	12#	0.189	0.109
PD6	0	113	0.179	0.038#
PD7	62*	66	0.138#	0.113*
PD8	92*	7#	0.258*	0.166*
PD9	95*	55	0.176	0.096
PD10	86*	25#	0.174	0.184*
PD11	90*	14#	0.330*	0.133*
PD12	100*	16#	0.160	0.207*
PD13	95*	34	0.158	0.087
PD14	48*	24#	0.138#	0.093
PD15	81*	70	0.192	0.165*
PD16	100*	6#	0.228*	0.197*
PD17	33*	55	0.157	0.113
FA1	0	7#	0.248*	0.075
FA2				
FA3	19	17#	0.302*	0.114*
FA4	70*	4#		
FA5	32*	8#	0.346*	0.081
FA6				
FA7	0	6#	0.352*	0.204*
FA8	0	15#	0.363*	0.034#
FA9	0	15#	0.297*	0.089
FA10	10	4#	0.353*	0.131*
FA11	8	5#	0.471*	0.101
CA1	48*	50	0.197	0.129*
CA2	19	36	0.202	0.140*
CA3	43*	18#	0.249*	0.071
CA4	76*	19#	0.252*	0.077
CA5	38*	34	0.193	0.088
CA6	57*	17#	0.235*	0.212*
CA7	62*	17#	0.301*	0.079
CA8	43*	50	0.203	0.161*
CA9	19	22#	0.260*	0.133*
CA10	25*	11#	0.318*	0.182*
CA11				
CA12	5	63	0.217	0.070
CA13	10	11#	0.476*	0.097
CA14	90*	12#	0.279*	0.071

Sub = subject; Inc = percentage of incomplete occlusions; Nmax = repetitions per train; Sylmed = median syllable duration; Sylvar = intra-train variation coefficient; NO = controls; HD, PD, FA, CA, see Table 1; bold numbers = upper (lower) limit of the normal range; *, # = above (below) normal range.

APPENDIX 2

Individual data on syllabic timing during production of articulation test sentences

Sub	N	Uttdur	s1	s2	s3	s4	s5	Smscan	Msvarc
NO1	24	0.912	0.168	0.140	0.225	0.214	0.159	0.922	0.093
NO2	24	0.810	0.145	0.135	0.211	0.191	0.133	0.911	0.126
NO3	24	0.911	0.178	0.138	0.222	0.233	0.146	0.894	0.125
NO4	24	0.915	0.192	0.138	0.245	0.194	0.143	0.891	0.101
NO5	24	0.883	0.164	0.139	0.223	0.191	0.166	0.938	0.136
NO6	24	0.764	0.141	**0.112**	0.210	**0.168**	**0.123**	0.879	0.126
NO7	24	0.899	0.170	0.134	0.243	0.229	0.126	**0.835**	0.141
NO8	24	0.793	0.137	0.126	0.207	0.176	0.144	0.920	0.111
NO9	24	0.815	0.151	0.139	0.194	0.176	0.146	**0.961**	0.099
NO10	24	1.052	**0.233**	0.153	0.243	0.285	0.155	0.856	0.093
NO11	24	**1.204**	0.221	**0.187**	**0.313**	**0.291**	**0.196**	0.896	0.097
NO12	24	0.855	0.159	0.134	0.231	0.191	0.139	0.898	0.107
NO13	24	1.078	0.218	0.163	0.260	0.259	0.174	0.911	**0.061**
NO14	24	0.964	0.180	0.129	0.249	0.235	0.171	0.873	**0.153**
NO15	24	**0.760**	**0.115**	0.126	**0.193**	0.186	0.134	0.893	0.106
HD1	24	0.778	0.113#	0.140	0.204	0.185	0.118#	0.866	0.181*
HD2	24	1.030	0.184	0.157	0.231	0.232	0.223*	0.944	0.129
HD3	24	1.375*	0.255*	0.213*	0.286	0.390*	0.271*	0.904	0.246*
HD4	24	1.179	0.211	0.171	0.261	0.212	0.271*	0.934	0.204*
HD5	24	1.293*	0.235*	0.193*	0.294	0.326*	0.239*	0.919	0.234*
HD6	23	0.894	0.171	0.127	0.228	0.211	0.141	0.882	0.167*
HD7	23	1.247*	0.224	0.229*	0.275	0.251	0.252*	0.986*	0.149
HD8	23	0.772	0.150	0.135	0.200	0.163	0.121#	0.928	0.263*
HD9	20	1.337*	0.280*	0.201*	0.308	0.224	0.305*	0.930	0.216*
HD10	22	1.578*	0.259*	0.260*	0.345*	0.340*	0.370*	0.945	0.267*
HD11	15	1.200	0.200	0.219*	0.298	0.207	0.298*	0.923	0.251*
HD12	11	1.397*	0.158	0.269*	0.225	0.254	0.533*	0.656#	0.240*
HD13	23	0.897	0.155	0.139	0.240	0.191	0.158	0.908	0.327*
HD14	0								
PD1	24	0.896	0.166	0.161	0.194	0.194	0.188	0.984*	0.117
PD2	22	0.661#	0.120	0.090#	0.173#	0.164#	0.131	0.877	0.242*
PD3	24	1.248*	0.255*	0.160	0.278	0.272	0.211*	0.903	0.373*
PD4	24	0.790	0.153	0.107	0.221	0.173	0.121#	0.846	0.197*
PD5	24	0.996	0.176	0.171	0.264	0.214	0.159	0.917	0.111
PD6	24	0.765	0.132	0.136	0.198	0.191	0.127	0.909	0.133
PD7	24	1.009	0.202	0.173	0.239	0.210	0.170	0.961	0.095
PD8	22	1.060	0.175	0.149	0.263	0.248	0.215*	0.894	0.514*
PD9	23	0.878	0.155	0.104#	0.216	0.251	0.137	0.782#	0.300*

APPENDIX 2

Continued.

Sub	N	Uttdur	s1	s2	s3	s4	s5	Smscan	Msvarc
PD10	21	0.782	0.130	0.149	0.198	0.158#	0.151	0.953	0.263*
PD11	24	1.013	0.206	0.171	0.243	0.217	0.185	0.964*	0.200*
PD12	23	0.826	0.202	0.124	0.163#	0.209	0.124	0.881	0.323*
PD13	19	1.060	0.171	0.209*	0.254	0.176	0.217*	0.948	0.365*
PD14	24	0.850	0.158	0.147	0.215	0.196	0.145	0.938	0.129
PD15	24	0.755#	0.124	0.111#	0.213	0.209	0.096#	0.758#	0.201*
PD16	12	2.577*	0.380*	0.807*	0.542*	0.279	0.384*	0.710#	0.295*
PD17	23	0.884	0.184	0.140	0.193	0.196	0.157	0.959	0.187*
FA1	24	1.087	0.129	0.176	0.304	0.223	0.291*	0.783	0.204*
FA2	24	0.942	0.188	0.161	0.233	0.221	0.171	0.950	0.158*
FA3	24	1.259*	0.245*	0.221*	0.244	0.259	0.313*	0.966*	0.200*
FA4	24	2.694*	0.499*	0.398*	0.444*	0.580*	0.534*	0.957	0.332*
FA5	24	1.565*	0.301*	0.271*	0.317*	0.313*	0.342*	0.986*	0.144
FA6	24	0.873	0.171	0.139	0.215	0.196	0.139	0.925	0.143
FA7	24	1.112	0.209	0.166	0.245	0.273	0.233*	0.934	0.135
FA8	24	1.384*	0.265*	0.245*	0.272	0.298*	0.293*	0.988*	0.156*
FA9	24	1.435*	0.317*	0.221*	0.284	0.326*	0.262*	0.953	0.162*
FA10	24	3.010*	0.372*	1.520*	0.473*	0.334*	0.260*	0.320	0.410*
FA11	24	1.795*	0.278*	0.373*	0.418*	0.370*	0.392*	0.955	0.166*
CA1	24	0.802	0.173	0.131	0.181#	0.194	0.123	0.923	0.205*
CA2	24	1.074	0.208	0.170	0.225	0.238	0.217*	0.969*	0.157*
CA3	24	1.373*	0.318*	0.242*	0.308	0.256	0.242*	0.964*	0.241*
CA4	24	1.237*	0.214	0.214	0.293	0.246	0.258*	0.965*	0.272*
CA5	24	0.960	0.190	0.167	0.205	0.204	0.183	0.986*	0.244*
CA6	24	1.245*	0.207	0.220	0.237	0.263	0.306*	0.952	0.269*
CA7	24	1.527*	0.322*	0.237*	0.292	0.394*	0.253*	0.920	0.197*
CA8	24	1.058	0.186	0.193	0.281	0.246	0.148	0.881	0.217*
CA9	24	1.469*	0.298*	0.262*	0.360*	0.289	0.265*	0.967*	0.209*
CA10	24	1.304*	0.291*	0.218	0.269	0.273	0.269*	0.977*	0.202*
CA11	24	1.512*	0.326*	0.216	0.346*	0.356*	0.291*	0.925	0.124
CA12	24	1.041	0.185	0.204	0.225	0.240	0.190	0.976*	0.170*
CA13	24	2.608*	0.514*	0.476*	0.657*	0.426*	0.509*	0.949	0.177*
CA14	24	2.210*	0.299*	0.613*	0.455*	0.320*	0.501*	0.832	0.138

Sub = subject; N = number of test sentences included; Uttdur = utterance duration; s1-s5 = median duration of the respective syllable across sentence repetitions; Smscan = scanning index; Msvarc = mean syllable variation coefficient.

APPENDIX 3

Individual data on intrasyllabic timing during production of articulation test sentences

Sub	VOTk	VOTp	VOTt	VOTlm	VOTdif	VOTvar
NO1	0.096	0.062	0.079	0.081	0.034	**0.112**
NO2	0.071	0.065	0.068	0.064	0.006	0.221
NO3	0.082	0.076	0.074	0.080	0.006	0.184
NO4	0.088	0.056	0.080	0.072	0.032	0.162
NO5	0.081	0.091	0.085	0.093	**-0.011**	0.128
NO6	0.087	0.076	0.075	0.080	0.010	0.172
NO7	0.093	0.061	0.078	0.076	0.032	0.191
NO8	0.066	0.058	0.061	0.062	0.008	0.140
NO9	0.071	0.046	0.058	0.058	0.025	0.138
NO10	0.070	0.070	0.064	0.063	0.001	0.169
NO11	**0.107**	**0.097**	**0.090**	**0.096**	0.010	0.165
NO12	0.099	0.049	0.057	0.069	**0.050**	0.180
NO13	0.084	0.052	0.054	0.062	0.033	0.175
NO14	0.092	0.067	0.078	0.080	0.025	0.138
NO15	**0.056**	**0.045**	**0.052**	**0.052**	0.011	**0.225**
HD1	0.084	0.064	0.074	0.075	0.020	0.166
HD2	0.068	0.048	0.047#	0.056	0.020	0.376*
HD3	0.065	0.055	0.067	0.057	0.010	0.319*
HD4	0.109	0.046	0.079	0.077	0.063#	0.373*
HD5	0.081	0.044#	0.060	0.057	0.037	0.303*
HD6	0.060	0.072	0.035#	0.051#	-0.012	0.356*
HD7	0.065	0.056	0.078	0.066	0.009	0.255*
HD8	0.058	0.049	0.056	0.053	0.009	0.336*
HD9	0.056	0.020#	0.031#	0.030#	0.036	0.486*
HD10	0.064	0.043#	0.033#	0.036#	0.021	0.693*
HD11	0.048#	0.032#	0.070	0.050#	0.016	0.303*
HD12	0.037#	0.038#	0.025#	0.028#	-0.001	0.554*
HD13	0.073	0.062	0.059	0.063	0.011	0.258*
HD14						
PD1	0.061	0.041#	0.056	0.053	0.020	0.288*
PD2	0.066	0.039#	0.038#	0.045#	0.026	0.233*
PD3	0.094	0.068	0.076	0.080	0.026	0.249*
PD4		0.028#	0.055	0.044#		0.214
PD5	0.100	0.094	0.077	0.085	0.006	0.157
PD6	0.066	0.039#	0.038#	0.048#	0.027	0.190
PD7	0.082	0.062	0.066	0.071	0.020	0.180
PD8	0.061	0.089				
PD9		0.056	0.075	0.059		0.286*

APPENDIX 3

Continued.

Sub	VOTk	VOTp	VOTt	VOT1m	VOTdif	VOTvar
PD10	0.072		0.067	0.062		0.255*
PD11	0.084		0.061	0.068		0.220
PD12		0.034$^{\#}$	0.007$^{\#}$	0.038$^{\#}$		0.328*
PD13	0.080	0.080	0.065	0.064	0.001	0.265*
PD14	0.087	0.077	0.086	0.087	0.010	0.148
PD15		0.050	0.062	0.062		0.176
PD16			0.148			
PD17	0.047$^{\#}$	0.020$^{\#}$	0.038$^{\#}$	0.029$^{\#}$	0.027	0.445*
FA1	0.093	0.084	0.102*	0.087	0.009	0.351*
FA2	0.077	0.076	0.066	0.065	0.001	0.354*
FA3	0.088	0.035$^{\#}$	0.042$^{\#}$	0.053	0.053*	0.445*
FA4	0.087	0.037$^{\#}$	0.025$^{\#}$	0.049$^{\#}$	0.050*	0.409*
FA5	0.095	0.040$^{\#}$	0.063	0.065	0.055*	0.373*
FA6	0.106	0.063	0.061	0.074	0.043	0.248*
FA7	0.071	0.033$^{\#}$	0.052	0.052	0.038	0.227*
FA8	0.051	0.040$^{\#}$	0.047$^{\#}$	0.043$^{\#}$	0.011	0.361*
FA9	0.093	0.051	0.063	0.062	0.042	0.450*
FA10	0.160*	0.114*	0.154*	0.147*	0.047	0.349*
FA11	0.113*	0.080	0.107*	0.100*	0.033	0.541*
CA1	0.066	0.048	0.049$^{\#}$	0.053	0.018	0.364*
CA2	0.080	0.052	0.069	0.071	0.028	0.317*
CA3	0.134*	0.119*	0.099	0.120*	0.015	0.254*
CA4	0.083	0.061	0.109*	0.082	0.022	0.294*
CA5	0.053$^{\#}$	0.032$^{\#}$	0.033$^{\#}$	0.039$^{\#}$	0.021	0.270*
CA6		0.081	0.075	0.086		0.215
CA7	0.070	0.034$^{\#}$	0.049$^{\#}$	0.046$^{\#}$	0.036	0.313*
CA8	0.088	0.056	0.053	0.063	0.032	0.264*
CA9	0.177*	0.112*	0.094	0.139*	0.065*	0.528*
CA10	0.130*	0.073	0.111*	0.109*	0.057*	0.204
CA11	0.116*	0.078	0.076	0.089	0.038	0.355*
CA12	0.074	0.066	0.069	0.067	0.008	0.235*
CA13	0.317*	0.215*	0.247*	0.250*	0.102*	0.243*
CA14	0.203*	0.094	0.133*	0.125*	0.109*	0.344*

Sub = subject; VOTk, VOTp, VOTt = median voice-onset-time of /k/, /p/ and /t/, respectively; VOT1m = median voice-onset-time across all three consonant categories; VOTdif = VOT difference between /k/ and /p/; VOTvar = variation coefficient of VOT.

APPENDIX 4

Invidual data on intrasyllabic timing during production of articulation test sentences

Sub	Vowel	Schwa	Cons1	Cons2	Vopart	Stress
NO1	0.107	0.072	0.154	0.100	0.418	0.606
NO2	0.090	0.053	0.166	0.099	0.365	0.617
NO3	0.108	**0.044**	0.184	0.127	0.337	0.626
NO4	0.086	0.067	0.178	0.104	0.366	0.615
NO5	0.092	0.059	0.164	0.098	0.370	0.619
NO6	**0.076**	0.049	0.155	0.093	0.342	0.624
NO7	0.117	0.055	0.177	0.113	0.375	0.641
NO8	0.082	0.059	0.145	0.093	0.391	0.601
NO9	0.092	0.047	0.144	0.082	0.399	0.652
NO10	**0.168**	0.072	0.156	0.129	0.444	0.608
NO11	0.126	0.046	**0.264**	**0.157**	**0.289**	**0.662**
NO12	0.103	0.068	0.159	0.088	0.416	0.633
NO13	0.119	0.064	0.190	0.140	0.375	0.609
NO14	0.104	**0.084**	0.177	0.124	0.394	**0.579**
NO15	0.109	0.073	**0.126**	**0.073**	**0.487**	0.632
HD1	0.066#	0.060	0.161	0.113	0.325	0.574#
HD2	0.131	0.093*	0.156	0.097	0.487	0.603
HD3	0.246*	0.121*	0.194	0.159*	0.475	0.606
HD4	0.123	0.117*	0.187	0.089	0.494*	0.577#
HD5	0.171*	0.111*	0.186	0.147	0.458	0.576#
HD6	0.109	0.070	0.146	0.098	0.451	0.617
HD7	0.111	0.078	0.195	0.145	0.371	0.585
HD8	0.083	0.069	0.141	0.078	0.415	0.630
HD9	0.114	0.115*	0.208	0.106	0.392	0.621
HD10	0.188*	0.160*	0.205	0.149	0.495*	0.561#
HD11	0.139	0.178*	0.143	0.060	0.576*	0.569#
HD12	0.158	0.118*	0.147	0.118	0.519*	0.530#
HD13	0.068#	0.059	0.194	0.119	0.290	0.581
HD14						
PD1	0.088	0.082	0.129	0.104	0.437	0.549#
PD2	0.093	0.082	0.096#	0.071#	0.521*	0.554#
PD3	0.152	0.102*	0.186	0.125	0.411	0.641
PD4	0.106	0.065	0.151	0.061	0.473	0.670*
PD5	0.101	0.083	0.168	0.112	0.422	0.595
PD6	0.114	0.051	0.143	0.078	0.431	0.669*
PD7	0.106	0.070	0.163	0.101	0.413	0.610
PD8	0.137	0.110*	0.161	0.113	0.483	0.577#
PD9	0.147	0.078	0.140	0.105	0.486	0.602

APPENDIX 4

Continued.

Sub	Vowel	Schwa	Cons1	Cons2	Vopart	Stress
PD10	0.100	0.069	0.136	0.059#	0.466	0.649
PD11	0.095	0.078	0.164	0.110	0.401	0.583
PD12	0.166	0.082	0.066#	0.043#	0.702*	0.634
PD13	0.097	0.086*	0.156	0.083	0.424	0.598
PD14	0.089	0.052	0.174	0.101	0.348	0.633
PD15	0.117	0.068	0.126	0.095	0.465	0.607
PD16	0.183*	0.166*	0.288*	0.128	0.445	0.641
PD17	0.110	0.094*	0.109#	0.082	0.529*	0.547#
FA1	0.110	0.123*	0.251	0.110	0.409	0.607
FA2	0.114	0.049	0.177	0.109	0.379	0.639
FA3	0.115	0.126*	0.165	0.140	0.437	0.531#
FA4	0.361*	0.287*	0.257	0.213*	0.575*	0.554#
FA5	0.180*	0.144*	0.235	0.143	0.463	0.585
FA6	0.081	0.042#	0.175	0.106	0.311	0.634
FA7	0.135	0.060	0.177	0.128	0.392	0.615
FA8	0.134	0.140*	0.188	0.149	0.444	0.526#
FA9	0.206*	0.113*	0.179	0.123	0.499*	0.608
FA10	0.139	0.115*	0.326*	0.196*	0.351	0.594
FA11	0.180*	0.135*	0.273*	0.169*	0.397	0.583
CA1	0.122	0.060	0.124#	0.076	0.483	0.639
CA2	0.144	0.087*	0.158	0.098	0.476	0.622
CA3	0.158	0.109*	0.229	0.108	0.433	0.654
CA4	0.130	0.092*	0.190	0.114	0.355	0.671*
CA5	0.113	0.097*	0.139	0.102	0.462	0.592
CA6	0.100	0.122*	0.166	0.150	0.410	0.512#
CA7	0.212*	0.126*	0.185	0.176*	0.466	0.571#
CA8	0.134	0.073	0.211	0.119	0.362	0.644
CA9	0.130	0.089*	0.248	0.163*	0.361	0.595
CA10	0.141	0.123*	0.219	0.128	0.429	0.589
CA11	0.183*	0.130*	0.254	0.171*	0.427	0.599
CA12	0.100	0.083	0.171	0.133	0.374	0.560#
CA13	0.187*	0.169*	0.490*	0.254*	0.308	0.619
CA14	0.176*	0.155*	0.317*	0.138	0.402	0.637

Sub = subject; Vowel = median duration of the target vowel; Schwa = median duration of post-target schwa; Cons1, Cons2 = median duration of C1, C2; Vopart = proportion of vocalic segments within the target word; Stress = durational stress contrast within the target word.

P. COPPENS, R.R. ROBEY

THE IMPORTANCE OF DYSARTHRIA IN DIFFERENTIAL DIAGNOSIS: A CASE STUDY

Dysarthria results from lesions in the central nervous system, the peripheral nervous system, or both. Therefore, dysarthria is associated with a wide variety of neurological pathologies and presents in a variety of forms. In a given clinical neurological picture, disturbance of speech is usually relegated towards the bottom of the list of symptoms; its presence may be acknowledged but granted little importance for diagnosis and usually for the rehabilitation process as well. The purpose of this chapter is to present a clinical case study illustrating the potential importance of dysarthria in the differential diagnosis of a neurological condition. Our patient was diagnosed with Binswanger's disease (BD), but the diagnosis did not match the dysarthria or some aspects of the clinical neuropsychological symptomatology. Subsequently, and in part due to the dysarthric signs, the diagnosis of amyotrophic lateral sclerosis (ALS) was added.

CASE STUDY: FIRST DIAGNOSIS

LS, a 71 year old female, was referred for outpatient speech-language pathology services on the basis of reduced intelligibility and swallowing difficulties. The patient carried the diagnosis of BD. In the nine months preceding her visit, the patient reportedly experienced two episodes of degradation in neurologic function. One of these episodes, presumably a TIA, occurred five months prior to her visit. At that time, the family reported that the patient's speech was slurred; however, her physician described her speech as normal and her language as relevant. An examination report two months later mentioned normal gait and normal cranial nerve functions. The report also mentioned a CT scan showing mild hyperdensities of the periventricular white matter that were more prominent in the frontal regions and reportedly "of doubtful clinical significance". No other lesions were observed and the sulci appeared normal.

Approximately one month later, that is, two months prior to her arrival in the speech-language pathology service, an MRI of the head was conducted. The T2 weighted axial images of the brain demonstrated multiple confluent high signal intensity foci of the periventricular white matter and of the centra semiovalia consistent with multiple deep white matter infarcts. The foci of increased signal intensity in the centra semiovalia are clearly visible in Figure 1 as extensive subcortical white areas. There was no evidence of

Y. Lebrun (ed.), From the Brain to the Mouth, 49–70.

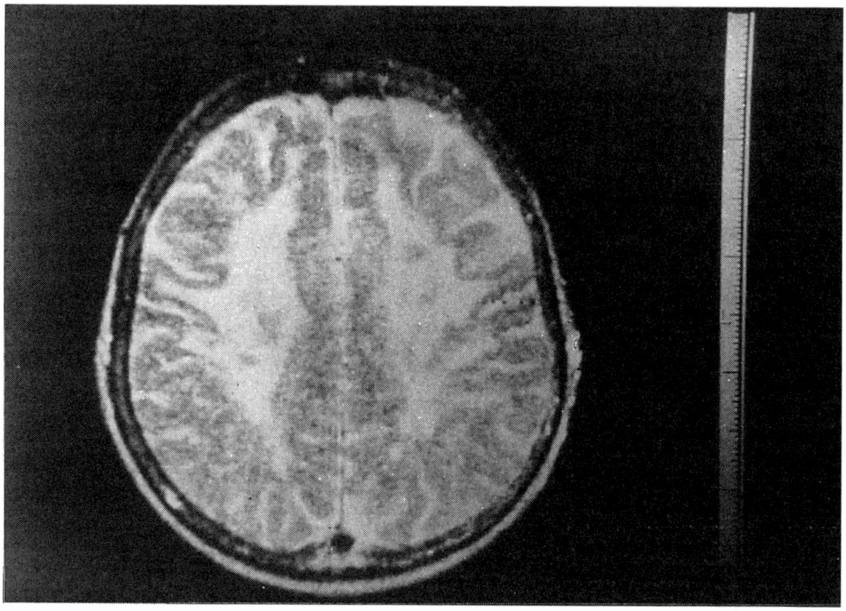

Fig. 1. LS's MRI showing extensive demyelination in the centra semiovalia.

neoplasm, subdural hematoma or communicating hydrocephalus. A slight lateral ventricle enlargement was noted. The diagnosis of BD was made at that time.

Binswanger's disease, or subcortical arteriosclerotic encephalopathy (Olszewski, 1962), is characterized by bilateral and symmetric demyelination of the periventricular white matter. Later in the course of the disease, the demyelination invades the entire centrum semiovale (Weisberg, Gerard, and Stasio, 1988). The cerebral cortex and the associated U-fibers are typically spared in BD. Lacunar infarcts in the basal ganglia and the pons may be present, but patients usually show no signs of brain stem infarction (Rosenberg, Kornfeld, Stovring, and Bicknell, 1979; Salomon, Yates, Burger, and Heinz, 1987). Moreover, there are no senile plaques or neurofibrillary tangles (De Reuck, Crevits, De Coster, Sieben, and vander Eecken, 1980).

Until recently, the diagnosis of BD could only be made confidently at autopsy. With the advent of modern imaging techniques, BD can be diagnosed in vivo (Ma, Lundberg, Lilja, and Olsson, 1992). Revesz, Hawkins, du Boulay, Bernard, and McDonald (1989) compared MRI scans with autopsy findings in four cases of BD and concluded that the MRI images directly corresponded to the histologic findings, thus suggesting a high validity for MRI technology as a diagnostic tool vis-à-vis BD. The clinical history of a patient suffering from BD is often positive for hypertension, strokes, or transient ischemic attacks. Clinical signs include a progressive dementia, personality changes, pseudobulbar affect, dysarthria, dysphagia, aphasia, incontinence, and gait problems (Caplan, 1995; Iijima, Ishino, Seno, Inagaki, and Haruki, 1993; Ma et al., 1992; Miller-Fisher, 1989; Santamaria Ortiz and Knight, 1994). Associated signs include bradykinesia and rigidity. Rarely described symptoms are hemianopia, anosognosia, and parietal neglect (Roman, 1987).

Regarding the neuropsychological signs associated with BD, impairments in memory and language are often reported (e.g., Bennett, Gilley, Lee, and Cochran, 1994; Bogucki, Janczewska, Koszewska, Chmielowski, and Szymanska, 1991). These signs are believed to be variable, but on average considered less important than those associated with other dementing illnesses such as Pick's or Alzheimer's diseases (Caplan, 1995). However, according to Santamaria Ortiz and Knight (1994), the "mental impairment was present in all cases, and was the main feature at onset in nearly 30%. It was present early in the illness in at least 60%" (p. 75). Classical aphasia syndromes have been reported, but they seem to be present only when there is a concomitant focal neurological lesion. The typical language impairments observed in most cases appear to be related to the dementing process.

Dysarthria is also common (Caplan, 1995; Rosenberg et al., 1979; Sacquegna, de Carolis, Daidone, and Dondi, 1988). Out of a sample of 81 patients from the literature (Bogucki et al., 1991; Burger, Gordon Burch, and Kunze, 1976; Caplan and Schoene, 1978; De Reuck et al., 1980; Derix, Hijdra, and Verbeeten, 1987; Dupuis, Brucher, and Gonsette, 1984; Garcia-Albea, Cabello, and Franch, 1987; Hershey, Modic, Jaffe, and Greenough, 1986; Hojer-Pedersen and Petersen, 1987; Iijima et al., 1993; Kosaka, Ikeda, Matsushita, and Iizuka, 1986; Ma et al., 1992; McQuinn and O'Leary, 1987; Rosenberg et al., 1979; Sako, Ito, Yasumura, Shimada, and Yokoo, 1991; Weisberg et al., 1988; Yamamura, Nishimura, Shirabe, and Fujita, 1987), we found that 36 were said to exhibit dysarthria or pseudobulbar palsy (usually defined as spastic dysarthria with emotional lability). This could nevertheless be an underestimation since "a terminal stage of pseudobulbar palsy (has) invariably occurred in 5 to 10 years" (Hojer-Pedersen and Petersen, 1987, p. 411). The characteristics of the

TABLE 1
Spastic Dysarthria Characteristics

1	Imprecise consonants
2	Monopitch
3	Reduced stress
4	Harsh voice quality
5	Monoloudness
6	Low pitch
7	Slow rate
8	Hypernasality
9	Strained-strangled quality
10	Short phrases
11	Distorted vowels
12	Pitch breaks
13	Breathy voice (continuous)
14	Excess and equal stress

motor speech disorder have never been reported in detail. However, the nature of the neurological process and the high frequency of pseudobulbar palsy leave little doubt that the dysarthria must be of the spastic type. The characteristics of spastic dysarthria as described by Darley, Aronson, and Brown (1975) are shown in Table 1. The rank order denotes the prominence level of specific signs.

In sum, a typical BD patient displays a progressive dementia (albeit sometimes subtle at onset) including neuropsychological and language signs, possible behavioral symptoms, and speech difficulties of the spastic type most often as part of a pseudobulbar palsy.

The oral-peripheral examination revealed that LS's face and mouth were symmetric at rest. However, when asked to smile, the patient would retract but not elevate the corners of her mouth. The patient's responses to requests for alternated smiling and puckering were slow with modest excursions. The patient did puff out her cheeks but lip closure appeared weak. At rest, the tongue was symmetric with fasciculations bilaterally. Range of motion for the tongue tip was fair, but more restricted on the right. Strength for pushing away the examiner's fingers on her cheeks was poor bilaterally. At rest, the velum was more elevated on the right. Contraction of the velum during sustained phonation of the vowel /a/ was more vigorous on the right. Nasal air emission was observed when she puffed out her checks, but could not be heard during phonation. A gag reflex was elicited on each side.

LS's speech sample and a reading of the Grandfather passage were characterized by a slow rate and poor intelligibility. Her conversation rate was 72.9 words per minute, and her reading rate 83.4 words per minute (normal ranges: 150-250 and 160-170 respectively, Yorkston, Beukelman, and Bell, 1988). Consonant precision was severely degraded. The weakness of articulatory contacts was very obvious, together with a reduced range of motion. There was spirantization of most plosives, probably the result of two mechanisms: a reduced range and strength of motion and velopharyngeal insufficiency. The velar plosives seemed proportionally more affected than the anterior stops. Final consonant deletion was a significant problem: 66.7% in conversation and 54.3% in the Grandfather passage. Figure 2 is a broad-band spectrogram of the word "pot" produced by LS. The deletion of the final consonant is apparent in this figure as is the absence of a burst-release for the initial plosive. For comparison purposes, Figure 3 is an analogous spectrogram of the same word produced by a normal speaker.

In addition, there were some instances of prolongation of phonemes, and prolonged intervals. Also, some vowels were neutralized. Figure 4 illustrates the vowel neutralization demonstrated by LS. The figure is a plot of the center frequencies of transitioning first and second formants in two productions of the diphthong [aI] (Kent et al., 1989). The diphthong was chosen for illustration because of the dynamic character of the formant structure typical for the production of this phone. Figure 5 displays the same diphthong uttered by a normal speaker using a broad-band spectrogram. A comparison of Figures 4 and 5 highlights the relatively little change in the formants in LS's two productions. The March 9 production, like many of LS's vowels, was considerably neutralized. Interestingly, the March 22 production shows greater second formant excursion than the March 9 production. However, the second portion of the diphthong still tends to be neutralized, as shown by the decreasing frequency of the second formant (Figure 4) instead of the normal increase (Figure 5).

The day to day variability of LS's productions was attributed to differences in prosody and affect. In the March 22 token, LS's production was characterized by self-initiated syllabic stress and emphasis which effected greater oral/lingual excursions than when her productions were less purposeful. Said differently, the degree of vowel neutralization, and therefore the degree of her intelligibility, was variable and that variability seemed

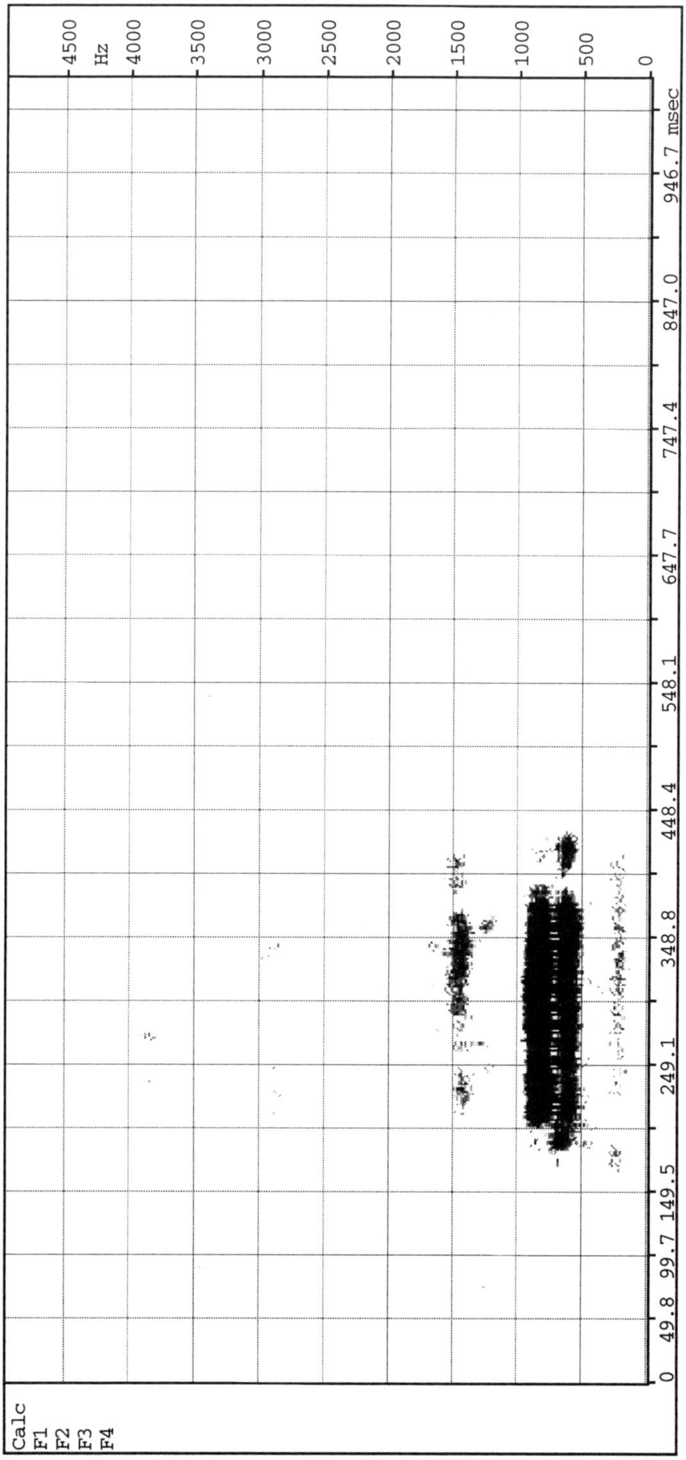

Fig. 2. Broad-band spectrogram of the word "pot" uttered by LS.

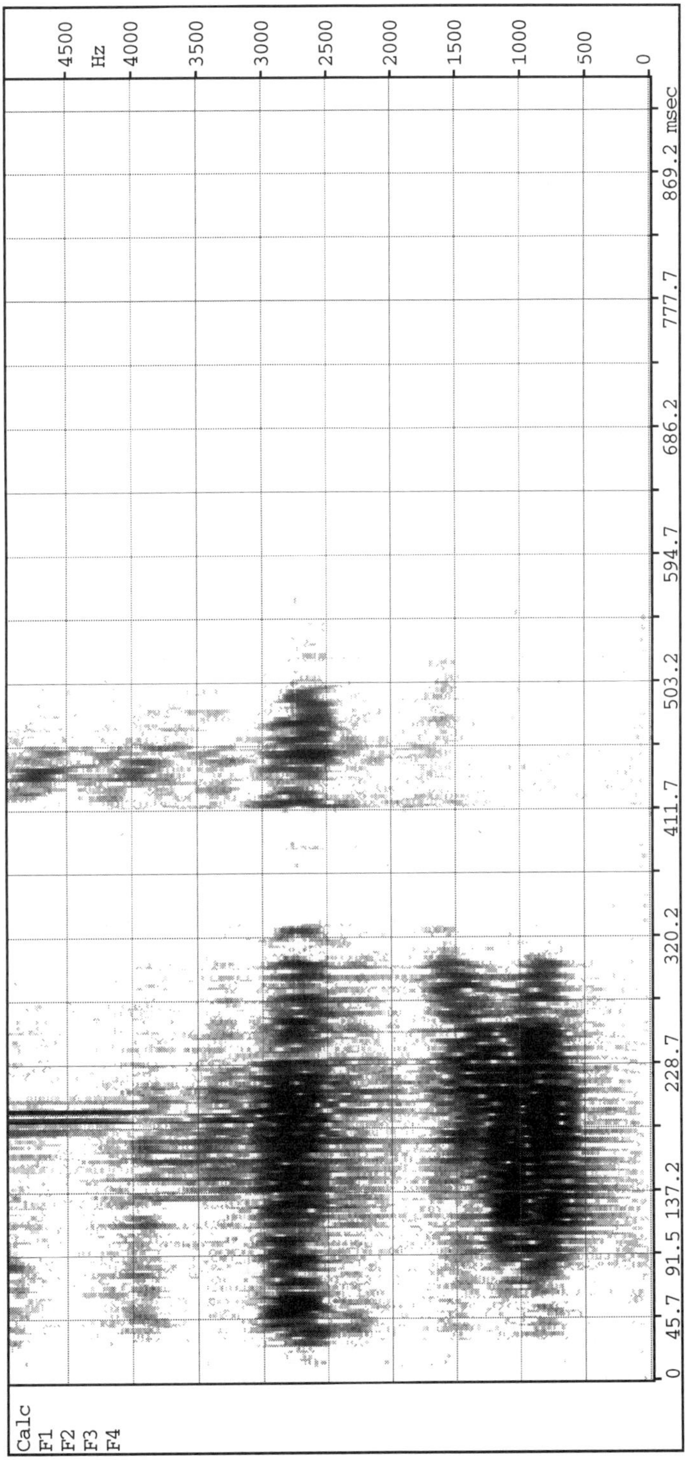

Fig. 3. Broad-band spectrogram of the word "pot" uttered by a normal speaker.

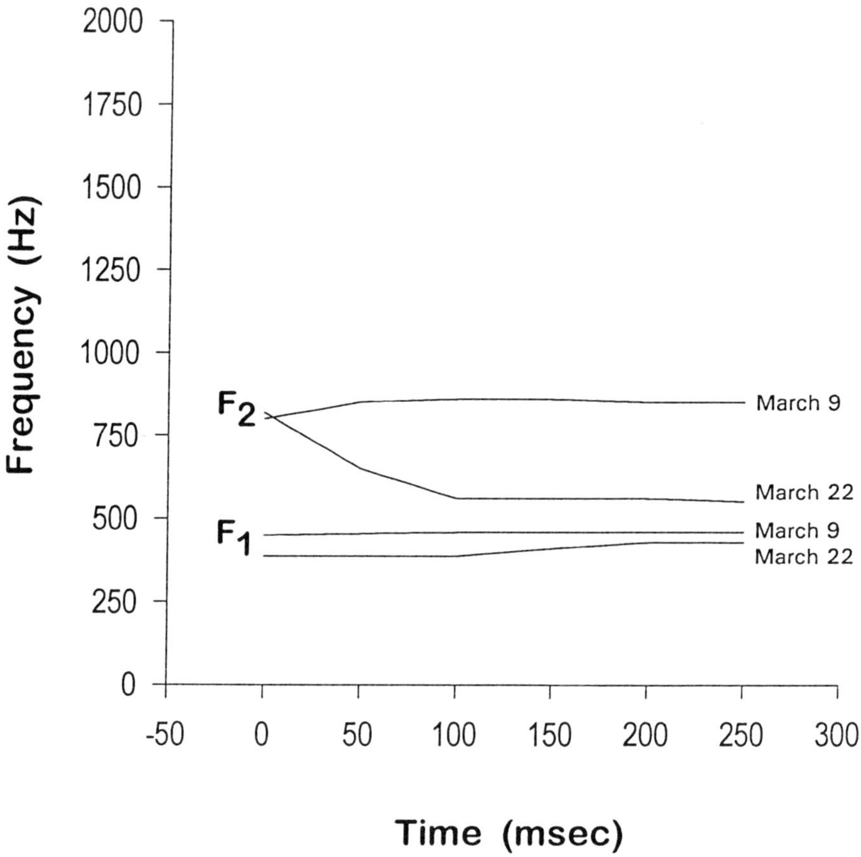

Fig. 4. F1 and F2 center frequencies for the diphthong [aI] produced in conversation by LS.

linked to the degree to which volitional effort and limbic input influenced her prosody.

Hypernasality was also a fairly prominent sign, particularly (but not exclusively) the assimilative hypernasality of vowels immediately preceding or following nasal consonants. However, no nasal escape was noted during conversational speech. Her voice quality was harsh with several instances of strained-strangled word initiations. Vocal loudness was low with only slight variations. Phonation was often accompanied by wet hoarseness. Vocal pitch was appropriate for age and gender although perceptually low and with little variation. Fatigue was also a factor. LS's articulation had a tendency to worsen with time within a session. LS's voice showed some

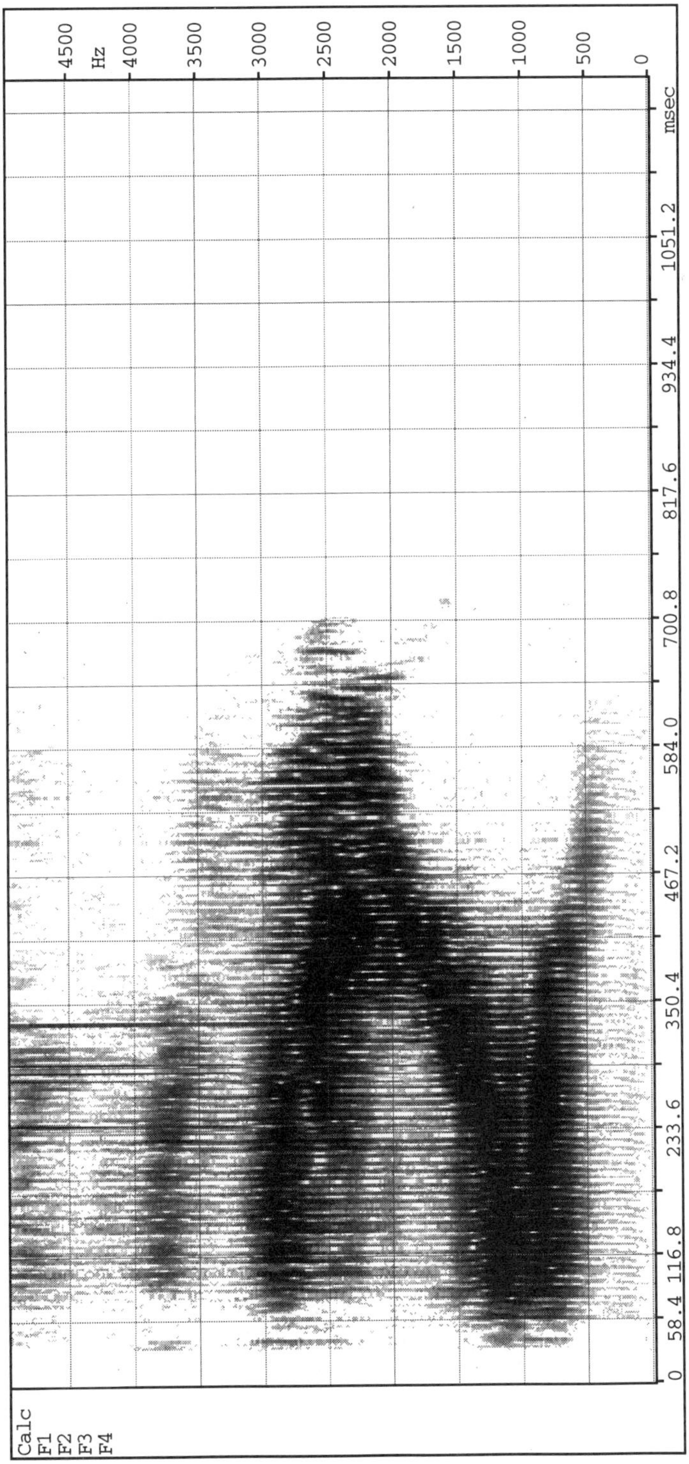

Fig. 5. Broad-band spectrogram of the diphthong [aI] produced by a normal speaker.

instances of excessive and equal syllabic stress. The harsh voice quality proved more evident in the sustained vowel task in which hyperfunction was obvious. She was able to prolong the vowel /a/ for 21 seconds, which is seven seconds more than the median value for normal elderly females (Duffy, 1995). The sustained vowel production did not demonstrate rhythmic or arrhythmic fluctuations of pitch. There were slight loudness variations, but there was no obvious nasality or breathiness.

Sequential and alternate productions of the consonants /p/, /t/ and /k/ were slow and imprecise. All productions were well below the minimum values for normal adults (Duffy, 1995). The accuracy of LS's productions was better for front versus back plosives. She uttered nine /pˆ / sequences in five seconds. The rhythm was regular, the vowels were slightly nasalized, and there was a harsh vocal quality. More than half the /p/ consonants were voiced and realized as /b/. Repeated productions of /tˆ / are displayed in a broad-band spectrogram in Figure 6. She produced eight /tˆ / syllables in five seconds. A burst release of the consonant is visible only in the initial production. No voice onset time is visible, supporting the perception of /dˆ / in this token. Contrary to the previous sequence, the irregularity of the timing of the productions is clearly visible as is a remarkable trailing-off effect. Perceptually, the patient's vocal quality was harsh, there was a slight spirantization of the consonant, and a slight nasal resonance. Figure 7 is a broad-band spectrogram of repeated productions of /kˆ /. LS produced seven /kˆ / syllables in sequence in five seconds. Notice that none of the productions contains a burst release for the plosive, essentially replacing it by a glottal fricative. As in Figure 6, timing was irregular and the last few productions were characterized by very little acoustic energy. A harsh voice quality was still present. Finally, a spectrogram of sequential productions of /pˆ tˆ kˆ / is found in Figure 8. LS produced three /pˆ tˆ kˆ / sequences in five seconds. In each sequence, the productions of /pˆ / demonstrate more acoustic energy than the productions of /tˆ /, which in turn demonstrate more acoustic energy than /kˆ /.

Moreover, acoustic energy diminishes progressively across productions of the sequences. Only the initial /pˆ / contains a burst release of the plosive. The low fundamental frequency of /tˆ / (fry) is evident in the vertical striations representing glottal pulses. Perceptually, the /k/ was always replaced by a glottal fricative, and the /t/ disappeared completely in two out of the three productions. The harsh voice quality and the slight hypernasality were still present. The rhythm was, once again, irregular. Although the distance between the /pˆ / and /tˆ / productions, and between the /tˆ / and /kˆ / productions is not as dyssynchronous as in ataxic dysarthria, they are

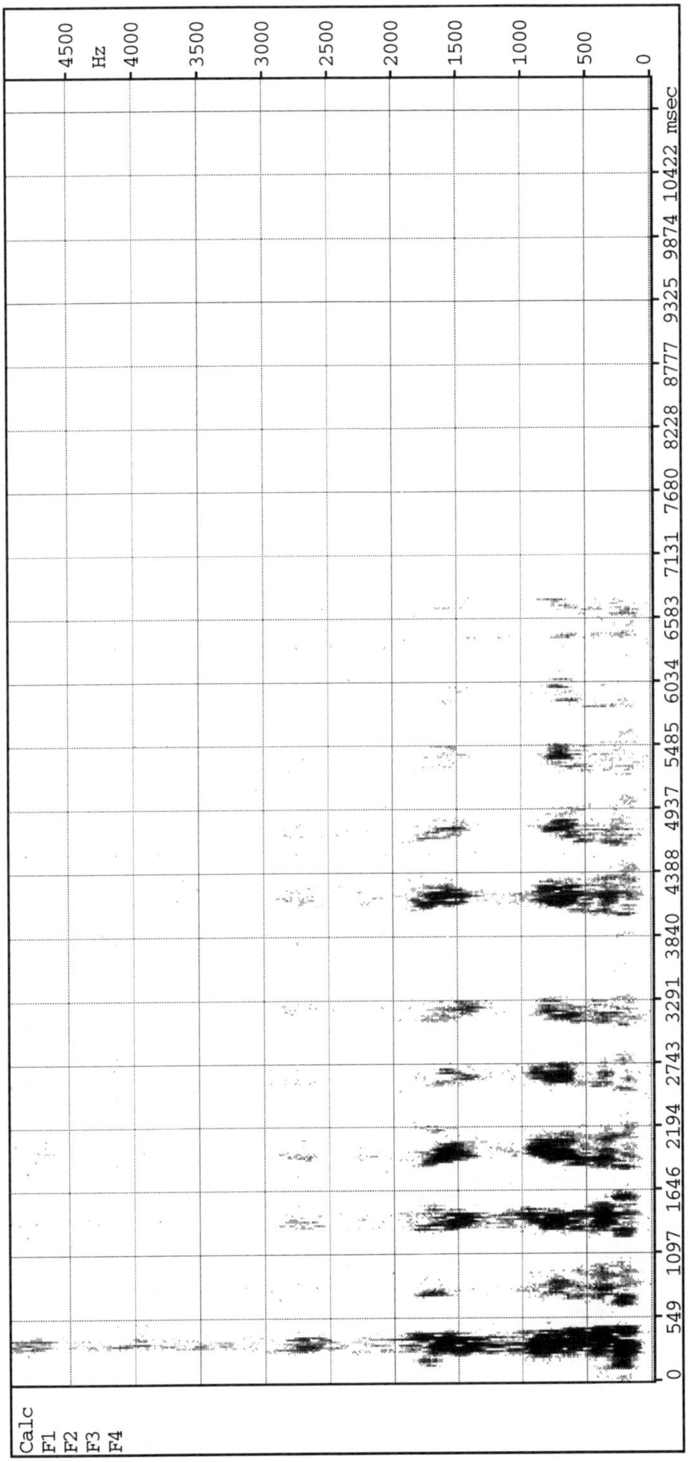

Fig. 6. Broad-band spectrogram of LS's repeated productions of /t'/.

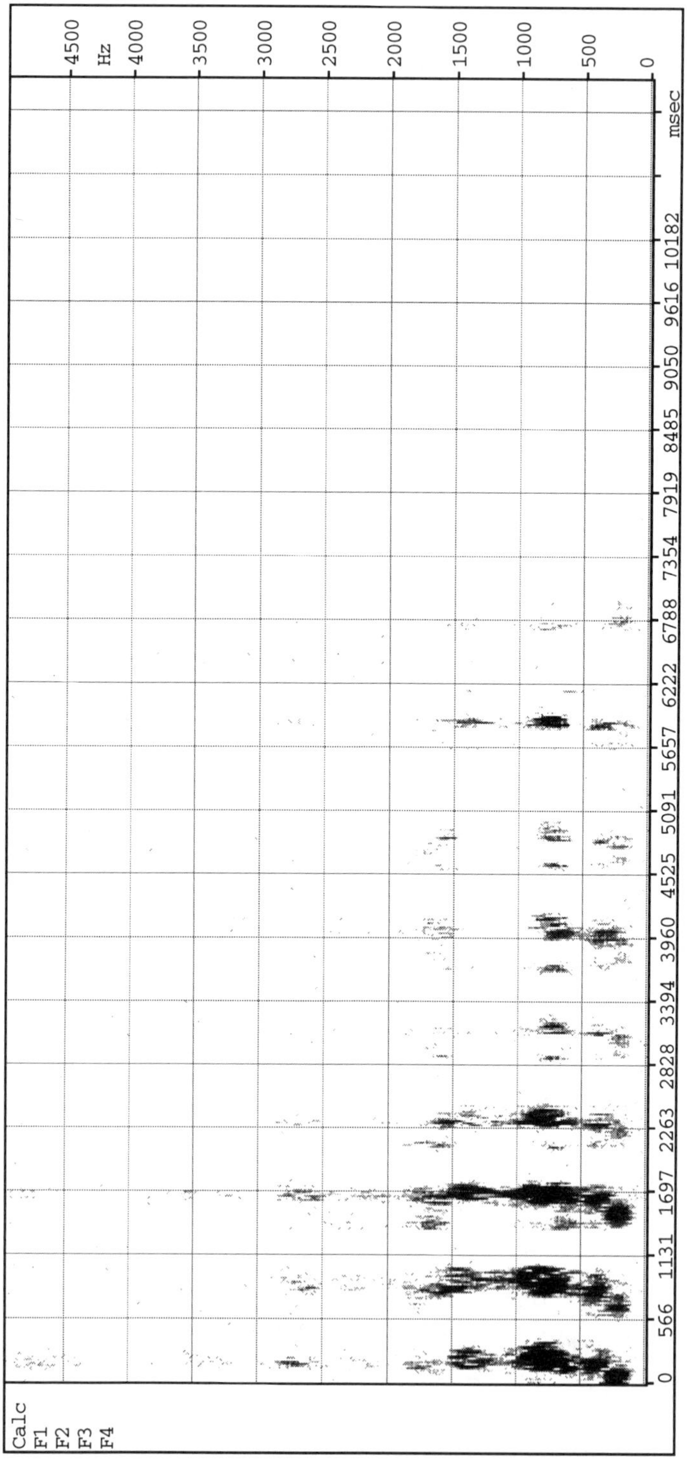

Fig. 7. Broad-band spectrogram of LS's repeated productions of /kˋ/.

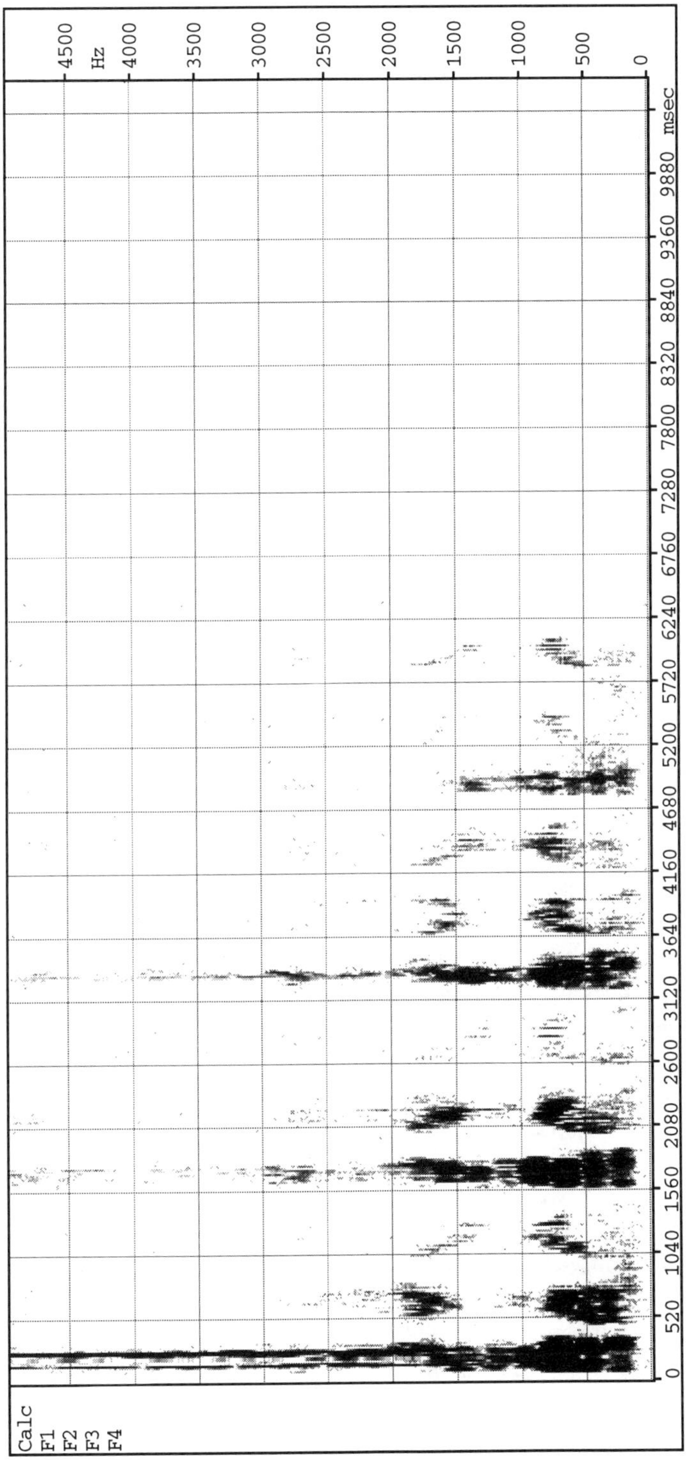

Fig. 8. Broad-band spectrogram of LS's repeated productions of /p`t`k`/.

not as regularly spaced as in normal productions. This characteristic led to a perceptually irregular rhythm.

The same task was repeated one week later, and the same results could be observed in terms of rhythm, slowness, and phoneme articulatory quality; however, her voice was stronger, the harsh quality had disappeared, and the slight hypernasality was no longer observed. It seems that these components were due to fatigue during the first evaluation. This confirms that fatigue was indeed a factor for LS.

A partial neuropsychological assessment was also performed. The patient could recall 4 out of 5 object names after a delay with distractor activity. Rey's Complex Figure Test yielded normal scores (copy: 50th percentile, memory: 25th percentile). Written calculations including the four basic operations, were successfully performed. The patient's performance on the Raven's Coloured Progressive Matrices was above the 75th percentile. She was oriented and appropriate in conversation at all times. The Boston Diagnostic Aphasia Examination was administered on two occasions. No language deficits were observed on either administration. Confrontation naming was adequate and her performance on the word fluency test (Borkowski, Benton, and Spreen, 1967) was within normal limits. In sum, the neuropsychological and language tests failed to identify any abnormalities. There was no sign of dementia. In addition, the case history or the patient interview did not reveal any behavioral or psychiatric disturbances. Emotional lability was not observed.

A conversation with LS two weeks later revealed substantial changes in her communicative function. Most of LS's utterances were now 1 to 6 word phrases. She frequently omitted articles and prepositions. Prolonged intervals increased in frequency. LS's speech appeared labored and was often produced in syllable-by-syllable fashion. In general, articulatory accuracy was severely degraded. Most vowels were neutralized. The differentiation of consonant productions was likewise reduced. Most of LS's consonant productions manifested as palatal glides or glottal stops. Many [+ continuant] consonants were simply continuations of the preceding vowels. With few exceptions, initial consonant clusters were reduced to singletons. Medial consonants were occasionally deleted. Final consonant deletion (singletons and clusters) was pervasive. Indeed, the entire final syllable was deleted in the production of polysyllabic words. Nasal assimilation was unchanged. The prevalence of these signs increased as the conversation progressed. Near the end of the interview, LS's productions consisted largely of neutralized vowels. The bilabial consonants and /h/ were most resistant to the progressively diminishing differentiation in the production of speech movements. Despite all this, LS occasionally demonstrated that she was capable of representing all of the phonetic constituents in a 1 to 2 word utterance. Vocal pitch was low, with little variation. The strained vocal quality now approximated that of a vocal fry. Audible exhalations were observed at the ends of many utterances.

The analysis of LS's dysarthria revealed that some characteristics were not easily reconciled with her diagnosis of BD. First, the spirantization of the plosives and the overall hypernasality suggested a level of velopharyngeal incompetence unusually severe for spastic dysarthria. Secondly, the diadochokinetic rates in spastic dysarthria are usually very regular (Duffy, 1995), whereas some of LS's productions showed irregular alternating motion rates. Particularly the former observation suggested that there might be a flaccid or lower motor neuron component to LS's pathology. This hypothesis was confirmed by the fact that tongue fasciculations were noticed during the oral-peripheral examination.

Equally surprising was the fact that LS showed no signs of dementia. The literature on BD indeed contains a number of patients reportedly without dementia (De Reuck et al., 1980; cases 2, 3, and 4; Hershey et al., 1986, cases 2 and 6; Huang, Wu, and Luo, 1985, cases 1, 2, 3, and 4; McQuinn and O'Leary, 1987). However, the absence of dementia might be related to the sensitivity of the tests used or one's definition of dementia. For example, McQuinn and O'Leary (1987) suggested that cognitive difficulties were not observed because the tests used were not sensitive enough, and Hershey et al. (1986) denied the presence of dementia but, on the other hand, acknowledged the presence of memory and attentional difficulties in one of their patients.

Similarly, BD has been reported to be asymptomatic in as much as a third of the cases (Kirkpatrick and Hayman, 1987; Pomeranz, 1989; Weisberg et al., 1988). It is probable that, because of the improvements in diagnostic imaging, many cases of BD are diagnosed much earlier in the disease course, and possibly before the cognitive impairments can be behaviorally observed or measured. If this were the case in LS, it would explain the negative results of the neuropsychological assessment. The same hypothesis could account for the absence of pseudobulbar affect in LS.

The conclusions of our assessment of LS were that her dysarthria showed unmistakable signs of spasticity, but that there were some concomitant characteristics of lower motor neuron involvement. The additional diagnosis of amyotrophic lateral sclerosis (ALS) was subsequently made by her attending physician.

TABLE 2
A Comparison of BD and ALS Symptoms

BD	ALS
Demyelination	Atrophy
Infarcts	
	Babinski Signs
Bradykinesia, Rigidity	
Paresis	Muscle Weakness
	Fasciculations
	Increased Muscle Tone
	Decreased Tendon Reflexes
Pseudobulbar Affect	Pseudobulbar Affect
Hypertension	
Hydrocephalus	
Gait Problems	
Incontinence	No Incontinence
Dementia	No Dementia
Unknown Dysarthria (probably spastic)	Mixed Dysarthria
Dysphagia	Dysphagia
Personality Changes	
Word Finding Difficulties	
Memory Problems	

CASE STUDY: SECOND DIAGNOSIS

Although the addition of the ALS diagnosis helped explain the dysarthria symptomatology, the next logical step was to try to understand the contribution of the two pathologies to the speech symptomatology. In other words, we knew that the patient's dysarthria was not typical for BD, but how typical was it for ALS?

ALS is a progressive degenerative process entailing cell loss and atrophy of both upper and lower motor neurons. The upper motor neuron involvement results in muscle weakness, increased muscle tone, hyperreflexia, Babinski signs, and pseudobulbar palsy. A degeneration of the pyramidal tracts, including the posterior limb of the internal capsule, and a loss of Betz cells are widely recognized. White matter lesions are occasionally observed on MRI scans, but the occurrence is less frequent than in the case of BD (Goodin, Rowley, and Olney, 1988). The lower motor neuron dysfunction includes muscle weakness and atrophy, fasciculations, and diminished or absent tendon reflexes. Usually, sphincter control is spared. Table 2 compares the motor, neurological, and neuropsychological signs of the two diseases.

Although ALS patients can show some subtle neuropsychological signs (Iwasaki, Kinoshita, Ikeda, Takamiya, and Shiojima, 1990), an associated dementia is highly exceptional (Mitsumoto, Hanson, and Chad, 1988; Poloni, Capitani, Mazzini, and Ceroni, 1986) and probably not related to ALS (see Yorkston et al., 1988). Dysphagia and dysarthria, however, are constant symptoms of ALS. The dysarthria observed in cases of ALS typically has both spastic and flaccid components. Early in the evolution of the disease, one of the two usually predominates. Later, both components are present and eventually the flaccid features become most noticeable because of the increasing muscle weakness (Yorkston et al., 1988). The six clusters of symptoms found to best characterize ALS are: prosodic excess, prosodic insufficiency, articulatory-resonatory incompetence, phonatory stenosis, phonatory incompetence, and resonatory incompetence (Darley, Aronson, and Brown, 1975). The first four of these symptom clusters represent the spastic aspect of ALS and the last two clusters are the consequence of the flaccid component of the disease (Duffy, 1995). The predominance of individual features will vary with the relative severity of the spastic and flaccid components in a given patient (Darley et al., 1988; Metter, 1985; Strand, Buder, Yorkston, and Ramig, 1994) and, in addition, an important within-subject measurement variability in patients with ALS is known to exist (Kent et al., 1991). Also, Darley et al. (1975) identified three signs that did not appear either with spastic or flaccid dysarthrias alone: prolongation of intervals, prolongation of phonemes, and inappropriate silences. These are believed to be triggered by the combination of spastic and flaccid dysarthrias in ALS. Table 3 compares the speech characteristics of spastic and ALS dysarthrias.

A more molecular analysis of ALS patients' speech reveals that the most disrupted features are velopharyngeal valving, lingual function, the slope of the second formant, final syllable deletion, cluster reduction, abnormal Fo, jitter and shimmer abnormalities, and harmonic-to-noise ratio (Kent et al., 1989; Kent et al., 1992; Mulligan et al., 1994).

LS did not display many of the signs believed to be pathognomonic for ALS dysarthria (i.e., prolongation of intervals, prolongation of phonemes, and inappropriate silences). Only two instances of prolongation of phonemes and one instance of prolonged interval were noted in LS's spontaneous speech and reading. This may indicate that in the case of our patient, the interaction between the two components may have been rather minimal.

Six weeks after the first examination, LS's spontaneous speech was less intelligible although her speaking rate had not significantly changed

TABLE 3
Comparison of Spastic and ALS Dysarthria

Spastic Dysarthria	ALS Dysarthria
1. Imprecise consonants	1. Imprecise consonants
2. Monopitch	2. Hypernasality
3. Reduced stress	3. Harshness
4. Harsh voice quality	4. Slow rate
5. Monoloudness	5. Monopitch
6. Low pitch	6. Short phrases
7. Slow rate	7. Distorted vowels
8. Hypernasality	8. Low pitch
9. Strained-strangled quality	9. Monoloudness
10. Short phrases	10. Excess and equal stress
11. Distorted vowels	11. Prolonged intervals
12. Pitch breaks	12. Reduced stress
13. Breathy voice (continuous)	13. Prolonged phonemes
14. Excess and equal stress	14. Strained-strangled quality
	15. Breathiness
	16. Audible inspiration
	17. Inappropriate silences
	18. Nasal emission

(71.2 words per minute). The same articulatory signs were observed, but with even more severe range of movements problem and more weakness. Monoloudness and monopitch were now noticed. Nasality was more prominent, but her voice was still characterized by harshness rather than breathiness (only in one alternate-motion-rate task some breathiness was identified, but not in conversation.) However, her spontaneous speech was now made of much shorter phrases than six weeks earlier. These two observations (no significant breathiness paired with shorter phrases) led us to conclude that the air wastage occurred at the velopharyngeal level rather than at the laryngeal level.

Short phrases are a frequent symptom in ALS dysarthria. Typically, in ALS, as in bulbar palsy, the short phrases are due to air wastage at the laryngeal level (i.e., breathiness) and that characteristic is part of the flaccid component of the mixed dysarthria. Because this was not the case in LS, we would argue that the flaccid component of LS's dysarthria was not as prominent as is generally the case in ALS dysarthria. Other signs support this hypothesis. For example, LS showed no audible inspirations, nasal emissions, or significant breathiness. Interestingly, she never displayed pseudobulbar affect, a symptom often associated with both pseudobulbar palsy and ALS.

In summary, if LS's dysarthria is compared to the usual clinical picture of ALS dysarthria, it seems that the flaccid signs are less obvious in our patient. In terms of the typical symptom clusters, articulatory-resonatory incompetence, phonatory stenosis, and resonatory incompetence were the most prominent. The last testing added elements of prosodic insufficiency. Prosodic excess and phonatory incompetence, which are also typical of ALS dysarthria, were absent or minimal. With only one of the clusters present which is usually associated with the flaccid component of ALS dysarthria (resonatory incompetence), it is no surprise that LS's speech was more spastic than expected. In addition, it also explains why hypernasality was the prominent sign of the flaccid component of her dysarthria.

CONCLUSIONS

There is no doubt that LS suffered from both BD and ALS. BD has been reported in association with Alzheimer's disease (Kosaka et al., 1986; Miller-Fisher, 1989), but to our knowledge, the patient presented here is the first description of a co-occurrence of BD and ALS.

Although the literature does not provide a clinical description of the dysarthria encountered in BD, a spastic dysarthria type, possibly as part of a pseudobulbar palsy (defined as spastic dysarthria and emotional lability), is a reasonable expectation. However, LS's dysarthria contained some signs of lower motor neuron pathology, and so, in a sense, appeared too flaccid for the diagnosis of BD. After the diagnosis of ALS was added, the flaccid component of her speech could be explained, but spasticity was undoubtedly the dominant characteristic. Compared to the typical symptomatology of dysarthria in ALS, the flaccid component of our patient's dysarthria was minimal. It is possible, if not probable, that LS's mixed spastic-flaccid dysarthria appeared more spastic than could be expected in ALS because of the presence of BD in addition to her ALS. Since this is the first description of a patient with both diagnoses, more information is needed to confirm that hypothesis. In the meantime, more detailed reports of the dysarthria symptomatology associated with BD are needed.

REFERENCES

BENNETT, D. A., GILLEY, D. W., LEE, S., COCHRAN, E. J. (1994). White matter changes: Neurobehavioral manifestations of Binswanger's disease and clinical correlates in Alzheimer's disease. *Dementia*, 5, 148-152.

BOGUCKI, A., JANCZEWSKA, E., KOSZEWSKA, I., CHMIELOWSKI, M., SZYMANSKA, R. (1991). Evaluation of dementia in subcortical arteriosclerotic encephalopathy

(Binswanger's disease). *European Archives of Psychiatry and Clinical Neuroscience*, 241, 91-97.

BORKOWSKI, J. G., BENTON, A. L., SPREEN, O. (1967). Word fluency and brain damage. *Neuropsychologia*, 5, 135-140.

BURGER, P. C., GORDON BURCH, J., KUNZE, U. (1976). Subcortical arteriosclerotic encephalopathy (Binswanger's disease). *Stroke*, 7, 626-631.

CAPLAN, L. R. (1995). Binswanger's disease revisited. *Neurology*, 45, 626-633.

CAPLAN, L., SCHOENE, W. (1978). Clinical features of subcortical arteriosclerotic encephalopathy (Binswanger's disease). *Neurology*, 28, 1206-1215.

DARLEY, F. L., ARONSON, A. E., BROWN, J. R. (1975). *Motor Speech Disorders*. Philadelphia: W.B. Saunders.

DE REUCK, J., CREVITS, L., DE COSTER, W., SIEBEN, G., VANDER EECKEN, H. (1980). Pathogenesis of Binswanger chronic progressive subcortical encephalopathy. *Neurology*, 30, 920-928.

DERIX, M. M. A., HIJDRA, A., VERBEETEN, B. W. J. (1987). Mental changes in subcortical arteriosclerotic encephalopathy. *Clinical Neurology and Neurosurgery*, 89 (2), 71-78.

DUFFY, J. R. (1995). *Motor Speech Disorders*. St Louis: Mosby.

DUPUIS, M., BRUCHER, J. M., GONSETTE, R. E., (1984). Observation anatomoclinique d'une encéphalopathie sous-corticale artérioscléreuse ("maladie de Binswanger") avec hypodensité de la substance blanche au scanner cérébral. *Acta Neurologica Belgica*, 84, 131-140

GARCIA-ALBEA, E., CABELLO, A., FRANCH, O. (1987). Subcortical arteriosclerotic encephalopathy (Binswanger's disease): A report of five patients. *Acta Neurologica Scandinavica*, 75, 295-303.

GOODIN, D. S., ROWLEY, H. A., OLNEY, R. K. (1988). Magnetic resonance imaging in amyotrophic lateral sclerosis. *Annals of Neurology*, 23, 418-420.

HERSHEY, L. A., MODIC, M. T., JAFFE, D. F., GREENOUGH, P. G. (1986). Natural history of the vascular dementias: a prospective study of seven cases. *Canadian Journal of the Neurological Sciences*, 13, 559-565.

HOJER-PEDERSEN, E., PETERSEN, O. F. (1987). Subcortical arteriosclerotic encephalopathy (Binswanger's disease). *Acta Medica Scandinavica*, 221, 409-412.

HUANG, K., WU, L., LUO, Y. (1985). Binswanger's disease: progressive subcortical encephalopathy or multi- infarct dementia? *Canadian Journal of the Neurological Sciences*, 12, 88-94.

IIJIMA, M., ISHINO, H., SENO, H., INAGAKI, T., HARUKI, S. (1993). An autopsy case of Binswanger's disease without hypertension and associated with cerebral infarction in the terminal stage. *The Japanese Journal of Psychiatry*, 47, 901-907.

IWASAKI, Y., KINOSHITA, M., IKEDA, K., TAKAMIYA, K., SHIOJIMA, T. (1990). Neuropsychological dysfunctions in amyotrophic lateral sclerosis: Relation to motor disability. *International Journal of Neuroscience*, 54, 191-195.

KENT, J. F., KENT, R. D., ROSENBEK, J. C., WEISMER, G., MARTIN, R., SUFIT, R., BROOKS, B. (1992). Quantitative description of the dysarthria in women with amyotrophic lateral sclerosis. *Journal of Speech and Hearing Research*, 35, 723-733.

KENT, R. D., KENT, J. F., WEISMER, G., MARTIN, R. E., SUFIT, R. L., BROOKS, B. R., ROSENBEK, J. C. (1989). Relationships between speech intelligibility and the slope of second-formant transitions in dysarthric subjects. *Clinical Linguistics and Phonetics*, 3, 347-358.

KENT, R. D., SUFIT, R. L., ROSENBEK, J. C., KENT, J. F., WEISMER, G., MARTIN, R. E., BROOKS, B. R. (1991). Speech deterioration in Amyotrophic Lateral Sclerosis: A case study. *Journal of Speech and Language Research*, 34, 1269-1275.

KIRKPATRICK, J. B., HAYMAN, L. A. (1987). White matter lesions in MR imaging of clinically healthy brains of elderly subjects: possible pathologic basis. *Radiology*, 162, 509-511.

KOSAKA, K., IKEDA, K., MATSUSHITA M., IIZUKA, R. (1986). A combination of Alzheimer's and Binswanger's diseases -a clinicopathological study of four cases-. *The Japanese Journal of Psychiatry and Neurology*, 40 (4), 685-692.

MA, K.-C., LUNDBERG, P. O., LILJA, A., OLSSON, Y. (1992). Binswanger's disease in the absence of chronic arterial hypertension. *Acta Neuropathologica*, 83, 434-439.

MCQUINN, B., O'LEARY, D. (1987). White matter lucencies on computed tomography, subacute arteriosclerotic encephalopathy (Binswanger's disease), and blood pressure. *Stroke*, 18, 900-905.

METTER, J. E. (1985). *Speech Disorders: Clinical Evaluation and Diagnosis*. New York: Spectrum Publications.

MILLER-FISHER, C. (1989). Binswanger's encephalopathy: a review. *Journal of Neurology*, 236, 65-79.

MITSUMOTO, H., HANSON, M. R., CHAD, D. A. (1988). Amyotrophic lateral sclerosis. Recent advances in pathogenesis and therapeutic trails. *Archives of Neurology*, 45, 189-202.

MULLIGAN, M., CARPENTER, J., RIDDEL, J., DELANEY, M. K., BADGER, G., KRUSINSKI, P., TANDAN, R. (1994). Intelligibility and the acoustic characteristics of speech in amyotrophic lateral sclerosis (ALS). *Journal of Speech and Hearing Research*, 37, 496-503.

OLSZEWSKI, J. (1962). Subcortical arteriosclerotic encephalopathy. Review of the literature on the so-called Binswanger's disease and presentation of two cases. *World Neurology*, 3, 359-375.

POLONI, M. CAPITANI, E., MAZZINI, L., CERONI, M. (1986). Neuropsychological measures in amyotrophic lateral sclerosis and their relationship with CT scan-assessed cerebral atrophy. *Acta Neurologica Scandinavica*, 74, 257-260.

POMERANZ, S. J. (1989). *Craniospinal Magnetic Resonance Imaging*. Philadelphia: W.B. Saunders.

REVESZ, T., HAWKINS, C. P., DU BOULAY, E. P. G. H., BARNARD, R. O., McDONALD, W. I. (1989). Pathological findings correlated with magnetic resonance imaging in subcortical arteriosclerotic encephalopathy (Binswanger's disease). *Journal of Neurology, Neurosurgery and Psychiatry*, 52, 1337-1344.

ROMAN, G. (1987). Senile dementia of the Binswanger type. *Journal of the American Medical Association*, 258, 1782-1788.

ROSENBERG, G., KORNFELD, M., STOVRING, J., BICKNELL, J. (1979). Subcortical arteriosclerotic encephalopathy (Binswanger): computerized tomography. *Neurology*, 29, 1102-1106.

SACQUEGNA, T., DE CAROLIS, P. DAIDONE, R., DONDI, S. (1988). Single-photon emission tomography with Technetium Tc 99m hexamethylpropylene amine oxime in Binswanger's disease (letter). *Archives of Neurology*, 45, 603-604.

SAKO, K., ITO, N., YASUMURA, S., SHIMADA, T., YOKOO, T. (1991). Binswanger's disease. Its clinical and magnetic resonance imaging (MRI) correlation. *Journal of Neurolinguistics*, 6, 25-38.

SALOMON, A., YATES, A., BURGER, P., HEINZ, R. (1987). Subcortical arteriosclerotic encephalopathy: brain stem findings with MR imaging. *Radiology*, 165, 625-629.

SANTAMARIA ORTIZ, J., KNIGHT, P. V. (1994). Review: Binswanger's disease, leukoaraiosis and dementia. *Age and Ageing*, 23, 75-81.

STRAND, E. A., BUDER, E. H., YORSTON, K. M., RAMIG, L. O. (1994). Differential phonatory characteristics of four women with amyotrophic lateral sclerosis. *Journal of Voice*, 8, 327-339.

WEISBERG, L., GERARD, G., STASIO, A. (1988). Computerized tomographic findings and differential diagnostic considerations in subcortical arteriosclerotic encephalopathy (Binswanger's disease). *Computerized Medical Imaging and Graphics*, 12 (4), 249-254.

YAMAMURA, T., NISHIMURA, M. SHIRABE, T., FUJITA, M. (1987). Subcortical vascular encephalopathy in a normotensive, young adult with premature baldness and spondylitis deformans. *Journal of the Neurological Sciences*, 78, 175-188.

YORKSTON, K. M., BEUKELMAN, D. R., BELL, K. R. (1988). *Clinical Management of Dysarthric Speakers*. Boston: College Hill.

J.-L. NESPOULOUS, N. MOREAU

REPAIR STRATEGIES AND CONSONANTAL CLUSTER PRODUCTION IN BROCA'S APHASIA

It was indeed the merit of Roman Jakobson to draw aphasiology away from the mere **surface description of symptoms** and to provide the first interpretations of aphasics' language disturbances in a **linguistically motivated** way (Jakobson, 1942). As far as segmental errors are concerned, he was thus able to interpret the nature of phonemic paraphasias by resorting to such linguistic theoretical constructs as **features** and **markedness** through which could be understood, for instance, the preferential tendencies often observed (on the paradigmatic axis) in phonemic substitutions, something that his predecessors — mainly clinicians — could not do! In the same way, although with less sophistication, (a) consonantal omissions within clusters were interpreted as **simplification of so-called "syntagmatic" patterns** and (b) some displacements of segments **as assimilations**, such as the ones philologists had been observing for years in diachronic studies.

Within the context of this paper, we will mainly concentrate on two error types frequently observed in aphasics' oral production[1] and commonly reported on in the aphasiological literature: **vocalic insertions and consonantal deletions in consonantal clusters**.

Contrary to most studies in such a field, though, our main goal will not be to focus upon the characterization of the underlying deficit and, thus, upon the nature of those specific processes supposed to be at fault in such and such patient, but will rather deal with the plausible "automatic" (?) coming into play, in the patients' oral production, of "repair strategies" they would mobilize whenever unable to compute and produce adequately target linguistic structures such as, here, consonantal clusters. If such repair strategies – as postulated by several modern phonological models (Pigott and Singh, 1985, Paradis, 1988) — do not only exist in "abstract" phonological theories but do also have some "psychological reality", as one used to say in early psycholinguistic research in the 60s, then, they should have (at least) the following properties:

(a) Since such strategies would be governed by general and powerful structural constraints which are part and parcel of the abstract, intrinsic and universal (?) properties of the speaker's "passive" phonological system, knowledge and competence, **they should generate the same symptoms at the surface level**, whatever might be the original causal factor(s) of the

[1] As well as in the oral production of many other speakers, e.g. first or second language learners.

Y. Lebrun (ed.), From the Brain to the Mouth, 71–80.

patient's deviant performance and pathology. More specifically, in the case of the production of consonantal clusters, all patients — be they aphasics, of one clinical type or another, or dysarthrics, of one type or another — should finally "repair" their production problem in a similar way before letting out their message.

(b) Ideally, since the word "strategy" usually implies the possibility of a choice between several alternatives, there should exist **several equivalent strategies to "repair" a same production problem**. Indeed, if there were only one single way to "repair" one specific production problem, one would only be allowed to speak of a "default repair process",[2] resulting systematically from a kind of **inertia of the linguistic system** enshrined in the speaking subject's mind/brain. Thus, when facing a specific difficulty — e.g. in producing consonantal clusters (Cf. infra) — a patient would resort either to a single preferential "repair strategy" or to several equivalent repair strategies alternatively.

The above clearly indicates that, within the context of this paper, we do consider pathological surface performance[3] as the potential coalescence of at least three factors that we can label:

– *positive factors* = what is still functioning in the patient's language processing system;

– *negative factors* = what is at fault at such and/or such a level in the same language processing system;

– and *adaptive* (palliative?) *strategies* whose aim is, so to speak, *to repair damage done to canonical linguistic representations and processes* ... in order to get the message through (Nespoulous, 1994).

We will mainly deal with the third type of factors in what follows.

I. VOCALIC INSERTIONS AND CONSONANTAL DELETIONS IN "FIRST GENERATION" LINGUISTICS AND APHASIOLOGY

Phonemic *insertions* and *deletions*, be they vocalic or consonantal, have always been present in classical philology as well as in first generation linguistics. Suffice it to mention here such terms as "epenthesis", "anaptyxis", "prothesis" and "intrusion" vs. "elision", "syncope", "apheresis" and "apocope", very frequently used, in particular by phonologists, etymologists and specialists in comparative linguistics. For the purpose of this paper, we will only retain two of these terms, whose interest has re-

[2]... and there are good reasons to believe that such "default repair processes", as here defined, do also exist and can be evidenced as well in aphasic oral performance.

[3]... and maybe "performance errors" in normals also ...

cently been renewed by several modern phonological theories (Cf. infra): *epenthesis* and *syncope*.

For first generation phonologists, an *epenthesis* is "a type of phonological intrusion in which an extra sound is inserted in a word".[4] In addition, epentheses are often subclassified into *anaptyxis* — if it is a vowel which is inserted between two consonants — and *prothesis* — if the inserted element is in word-initial position. In contrast, a *syncope* consists in the deletion of a (most often) atonic vowel within a word. It is called *apheresis* if it is the initial sound of a word which is deleted and *apocope* if the deletion involves the final element.

However, if such labels are indeed useful to identify different surface manifestations within and across languages, they mainly remain *descriptive tools* without any explanatory value from a theoretical point of view. Only occasionally does one find mention of the fact that both epentheses and syncopes tend (a) to simplify or shorten phonemic strings in free running connected speech and/or (b) to improve the message euphony.

It is with such descriptive tools that first generation linguists, psycholinguists and neurolinguists attempted to account for some phonemic and/or phonetic deviations frequently observed in aphasics' language disturbances as well as for those that could be gathered from child language acquisition (Jakobson, 1942). For them all (see Alajouanine, Ombredane and Durand, 1939), no doubt following Jakobson and Troubetzkoy's markedness theory (1939/1970), such error types aimed at simplifying the verbal output. When applied specifically to the production of consonantal clusters, such a framework made it possible to understand (a) why such "complex" phonemic structures were "learnt" and mastered fairly late by children in the acquisition of their phonological system and (b) why they were frequently disturbed and "destroyed" in the speech of many an aphasic patient (Blumstein, 1973, Puel, Nespoulous, Bonafé and Rascol, 1980, Kilani-Schoch, 1982, Martory and Messerli, 1983, Nespoulous, Joanette, Béland, Caplan and Lecours, 1984). Yet, one had to wait until fairly recently to take a further and crucial step forward in the understanding of the underlying determinism of such phenomena as vocalic insertions and consonantal deletions in consonantal cluster production.

[4] From "International Encyclopedia of Linguistics", Oxford University Press, 1992.

II. VOCALIC INSERTIONS, CONSONANTAL DELETIONS IN CONSONANTAL CLUSTERS: THE NEED FOR SYLLABLES AND SYLLABIC STRUCTURES WITHIN PHONOLOGICAL THEORY

In first generation phonology, there was little concern for syllables in the description of the structural properties of a natural language. Indeed, linguists often limited their work to the identification of the different phonemes and features that were "relevant" within the phonological inventory of a given language (as opposed to another) and they limited their characterization of the syntagmatic, or phonotactic, patterns to the **linear description of phonemic combinations** that were acceptable or not in that language. Occasionally, though, they would go a little further in their description of phonemic sequences by differentiating (a) "open syllables" — of the /C + V/ type — from "closed syllables" — of the /C + V + C/ type — and (b) "stressed syllables" from "unstressed syllables", in those languages where the placement of stress did affect the lexical meaning of a word. In fact, for a number of decades, syllables seemed to interest more phoneticians than phonologists. A good example, among many others, is certainly Whitney's "sonority principle"[5] which was first stated as early as 1865 (!), then used by prominent phoneticians such as Sievers (1881) and Jespersen (1909) but that was only integrated within phonological models very recently (Lekach, 1979; Selkirk, 1982; Clements, 1990).

As far as consonantal clusters are concerned, they were, without further comments, mainly considered as increasing the internal complexity of a syllable. Thus, the first aphasiologists — observing that patients often tended to "destroy" consonantal clusters — interpreted, in very general terms, such cluster destructions as the outcome of a *simplification process*, just as in many other pathological manifestations.[6] In no way, though, did they analyze more in depth and did they interpret more accurately such phenomena as vocalic insertion and consonantal deletion ... which nevertheless often led to such cluster destructions. But things did begin to change with the fairly recent developments of "metrical phonology" (Prince, 1983, Halle and Vergnaud, 1987).

According to this new trend in phonology, phonemic strings are represented in a hierarchical manner, much in the same way as syntactic structures have been in most, if not all, modern syntactic models and theories. "Originally introduced as a hierarchical theory of stress, the approach now

[5]According to such a principle, the different segments that constitute a syllable are arranged in a sequence of first increasing then decreasing sonority.

[6]e.g. at the morphological and syntactic levels.

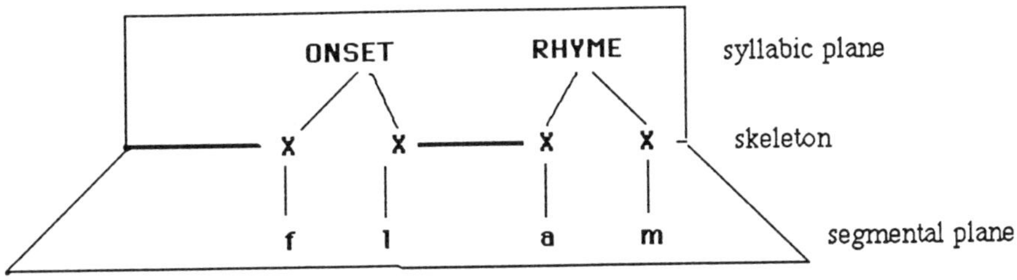

Fig. 1.

covers the whole domain of syllable structure".[7] A syllable thus "consists of an obligatory nucleus[8] with optional initial and final margins".[9] In cases in which a syllable does have margins, it is represented as a hierarchical structure whose superior node dominates two constituents: the "onset" and the "rhyme", the latter being further subdivided into "peak" (or "nucleus") and "coda". The onset corresponds to the consonant(s) which precede(s) the vocalic nucleus, the rhyme corresponding to the vocalic nucleus and the following consonant(s). Whenever syllable margins are made of several consonants, they are said to be "branching" (= branching onsets or branching codas).

The appearance of the syllable in phonological theory is closely linked to the development of a non-linear conception of phonological representations. Nowadays, such representations are thus often described as tridimensional objects with (a) the "syllabic" plane, (b) the "segmental" plane, (c) linked together by the "skeleton". Within the context of some theories, the "skeleton" is characterized as a "CV type" sequence (Clements and Keyser, 1983) whereas, in others, it is only — as represented in Figure 1 — a series of "X" corresponding to "positions" or "time units" without phonological content.

As far as consonantal clusters are concerned, they are considered as "branching" constituents complexifying one syllable margin or the other, either the syllable onset ("branching onset") or the syllabic coda ("branching coda"). They thus constitute more marked syllabic structures than the standard unmarked /CV/ type syllable (Kaye and Lowenstamm, 1984).

Applied to aphasic data, such modern phonological models make it pos-

[7] From *International Encyclopedia of Linguistics*, Oxford University Press, 1992.

[8] The nucleus is most of the time a vowel but liquids and nasals may also be syllabic nuclei.

[9] From *International Encyclopedia of Linguistics*, Oxford University Press, 1992.

sible to go definitely further in the understanding of segmental errors to be observed in the (mis)production of consonantal clusters. Following Blumstein (1978) and Béland (1985), Valdois (1988) is certainly one of the very first to have analyzed in-depth addition and omission errors "produced by aphasic subjects in the context of two-consonant clusters" resorting to such models. Convinced of the important limitations of former analyses of errors on consonantal clusters produced by either non-brain-damaged[10] or aphasic subjects, she tried and assessed the vulnerability of both consonants involved in CC clusters. According to her, (a) the onset-rhyme distinction was not sufficient to characterize syllabic structure because (b) both consonants present in a cluster might not have the same "representational weight". On the basis of 243 cluster "destructions" (involving the omission of a consonant) and of 130 cluster "creations" (involving the addition of a consonant), she was thus able to demonstrate that:

– clusters of the "/s/ + consonant" type behaved differently from other types of consonantal clusters;

– C2 (i.e. the second consonant of a cluster) was preferentially "disturbed" (= omitted, mainly) when the error involved a cluster of the "obstruent-liquid" type (e.g. /tr/→/t/);

– consonantal additions mainly appeared in C2 position, the initial consonant (or C1) being always an obstruent in such cases (/p/, /t/, /k/...);

– additions in C1 position always occurred when C2 was a liquid (e.g. /r/→/tr/).

She thus clearly emphasized (a) the importance of the internal (= segmental) structure of two consonant clusters and (b) the necessity of considering the existence of an interacting effect between the segmental level and the syllabic level if we are to understand ever the underlying determinism of both cluster destructions and cluster creations.

III. CLUSTER DESTRUCTION, VOCALIC EPENTHESIS AND CONSONANTAL SYNCOPE

Béland (1990) mainly worked on vocalic epentheses and syncopes. Valdois (1990) mainly worked on consonantal additions and omissions. Our main aim in what follows is to consider as two equivalent repair strategies vocalic epenthesis and consonantal syncope in cluster destruction or "reparation", both processes leading to the production of an unmarked syllabic structure of the /CV/ type.

[10]See Fromkin (1973), Shattuck-Hufnagel and Klatt (1979), Kupin (1982), Stemberger (1984), Laubstein (1985), Stemberger and Treiman (1986).

TABLE 1

Percentages of destruction and creation of consonantal clusters for subjects with Broca's aphasia and with conduction aphasia

	N	Destruction	Creation
Broca 1	83	62.6%	37.4%
Broca 2	95	67.5%	32.4%
Broca 3	77	92.2%	7.8%
Broca 4	57	54.3%	45.7%
Conduction 1	137	53.3%	46.7%
Conduction 2	82	62.1%	37.9%
Conduction 3	33	51.5%	48.5%
Conduction 4	94	47.8%	52.2%

TABLE 2

Percentages of destruction of consonantal clusters by syncope Vs. by epenthesis

	Syncope	Epenthesis
Broca 1	38%	62%
Broca 2	42%	58%
Broca 3	88%	12%
Broca 4	29%	71%

In order to test such an hypothesis, we resorted to data published by Nespoulous et al. (1984). In that investigation — whose aim was to document the existence of a "markedness effect" in the aphasic production of consonantal clusters (i.e. CCV sequences being replaced by CV sequences far more often than the reverse) — it was found (see Table 1) that such a tendency was only clearly present in patients with a phonetic impairment (= Broca's aphasics) and absent in patients without such an impairment (= conduction aphasics).[11] At the time, though, we did not resort to any of the new advances in phonological theory we have just mentioned.

If we re-examine — with more sophisticated tools and particularly with the notion of "repair strategy" in mind (Cf. supra) — the error patterns of the patients that were frequently "destroying" clusters, i.e. Broca's aphasics, we observe (see Table 2):

– that one of them (Broca 3) is resorting to syncope (or deletion) of the

[11] Indeed, in most cases, conduction aphasics equally created or destroyed consonantal clusters. Nevertheless, when destroying such clusters, these patients resorted equally to syncope or to epenthesis.

second consonant (C2) of a cluster (C1C2) in 88% of the cases and that he is only resorting to epenthesis in the remaining 12% of the cases; ... by the way, such a patient clearly corresponds to the most impaired subject (out of the 4) on classical clinical criteria.

– that a second patient (Broca 4) — who is the only patient with phonetic impairment who produces almost as many cluster creations (45,7%) as cluster destructions (54,3%) — more frequently resorts to epentheses than to syncopes (71% as opposed to 29%).

– that the other two patients (Broca 1 and 2) resort to epentheses more often than to syncopes but at a lesser (although statistically significant) degree than the previous patient: 62% vs. 38% in Broca 1; 58% vs. 42% in Broca 2).

Moreover, in all four patients and in all cases, syncopes systematically affect:

– the liquid that is in C2 position in a /plosive + liquid/ consonantal cluster; e.g.: /pR/→/p/.

– the liquid that is in C2 position in a /fricative + liquid/ consonantal cluster; e.g.: /fR/→/f/.

– and the fricative that is in C1 position in a /fricative + plosive/ CC; e.g.: /sp/→/p/.

As far as epentheses are concerned, they systematically consist in the insertion of either a schwa or a vowel already present in the target word in between a plosive (or a fricative) and a liquid, no vocalic insertion between a fricative and a plosive being observed (in clusters such as /st/ or /sb/ ...).

What such data suggest is:

(a) that, linguistically speaking and as clearly stated by Béland in her doctoral dissertation (1985), both epenthesis and syncope indeed are equivalent "repair strategies" available (considering the structural phonological properties of a language such as French) to a speaking-subject who, for whatever reason, is having difficulties with the production of consonantal clusters. Interestingly enough, Broca 2 even resorts to both strategies — epentheses and syncopes — at different moments on the same lexical items, which is certainly a clear indication of the equivalence of both strategies.

(b) that, psycholinguistically speaking, it is quite possible for a patient to resort, either preferentially or alternatively, to one of the two equivalent above-mentioned strategies (whether consciously or not) when facing a production problem, even if the latter stands at a low (peripheral) level of processing.

If our interpretation is correct, it thus indicates that the characterization of segmental error patterns in terms of segment addition or deletion, at least in the cases mentioned above, must be more accurately accounted for in terms of the coming into play of "repair strategies" predictable from the structural properties of the natural language spoken by the patient (and maybe also predictable from universal structural properties of phonological systems in general) than in the terms provided by any "error generation model" of the type used in first generation aphasiology. If so, it would be a clear indication of the role played by high-level, abstract, universal (?) phonological constraints on low(er)-level phonetic or motor processes, regardless of the fact that the observation be done (a) within the context of aphasia following cortical brain damage, as in Broca's aphasia, (b) within the context of subcortical dysarthrias (an observation that remains to be accurately assessed), (c) within the context of developmental disorders or dysphasias or even (d) within the context of "normal" language acquisition, when phonetic skills have not quite (or yet) been acquired.

REFERENCES

ALAJOUANINE, T., OMBREDANE, A., DURAND, M. (1939) *Le syndrome de désintégration phonétique dans l'aphasie*, Paris, Masson.

BÉLAND, R. (1985) Contraintes syllabiques sur les erreurs phonologiques dans l'aphasie, Thèse pour le Doctorat, Université de Montréal.

BÉLAND, R. (1990) Vowel epenthesis in aphasia, in: Nespoulous J- L., Villiard P. (Eds.) *Morphology, phonology and aphasia*, New York, Springer Verlag: 235-252.

BLUMSTEIN, S. (1973) *A phonological investigation of aphasic speech*, Mouton, The Hague.

BLUMSTEIN, S. (1978) Segment structure and the syllable in aphasia, in Bell A., Hooper J. (Eds.) *Syllables and segments*, Amsterdam, North Holland: 189-200.

CLEMENTS, G. (1990) The role of the sonority cycle in core syllabification, in Kingston J., Beckman M. (Eds.) Papers from the first conference on *Laboratory phonology*, Cambridge and New York, Cambridge University Press: 283-333.

CLEMENTS, G., KEYSER, S. (1983) CV phonology: a generative theory of the syllable, *Linguistic Inquiry Monographs*, 9., Cambridge, Mass., The M.I.T Press.

FROMKIN, V. (1973) *Speech errors as linguistic evidence*, The Hague, Mouton.

HALLE, M, VERGNAUD, J. (1987) An essay on stress, *Current Studies in Linguistics*, 15, Cambridge, Mass., The M.I.T Press.

JAKOBSON, R. (1942) *Kindersprache, Aphasie und allgemeine Lautgesetze*, Uppsala Univ. Arsskr., 9.

JESPERSEN, O. (1909) *A modern English grammar on historical principles*, vol. 1., *Sounds and spelling*, Heidelberg, Winter.

KAYE, J., LOWENSTAMM, J. (1984) De la syllabicité, in Dell F., Hirst D., Vergnaud J. (Eds.) *Forme sonore du langage*, Paris, Hermann. :123-161.

KILANI-SCHOCH, M. (1982) Processus phonologiques, processus morphologiques et lapsus dans un corpus d'aphasique, Thèse pour le Doctorat, Berne, Publications Universitaires Européennes.

KUPIN, J. (1982) *Tongue twisters as a source of information about speech production*, Bloomington, Indiana University Linguistic Club.

LAUBSTEIN, A. (1985) *The nature of production grammar syllable*, PhD Dissertation, University of Ottawa.

LEKACH, A. (1979) Phonological markedness and the sonority hierarchy, in Safir K. (Ed.) Papers on syllable structure, metrical structure and harmony processes, *MIT Working Papers in Linguistics*, 1, Cambridge, Mass.: 172-177.

MARTORY, M., MESSERLI, P. (1983) Analyse comparée des troubles de l'expression orale, in Messerli, P., Lavorel, P.M., Nespoulous, J-L (Eds.) *Neuropsychologie de l'expression orale*, Lyon, Editions du CNRS: 71-91.

NESPOULOUS, J-L. (1994) Linguistique, neurolinguistique et neuropsycholinguistique. Un parcours en quatre étapes, in Séron, X., Jeannerod, M. (Eds.) *Neuropsychologie humaine*, Bruxelles, Mardaga: 317-319.

NESPOULOUS, J-L., JOANETTE, Y., BÉLAND, R., CAPLAN, D., LECOURS, A.R. (1984) Phonologic disturbances in aphasia: is there a "markedness effect" in aphasic phonetic errors? *Advances in Neurology*, 42: 203-213.

PARADIS, C. (1988) On constraints and repair strategies, *The Linguistic Review*, 6.1.: 71-97.

PIGGOTT, G., SINGH, R. (1985) The phonology of epenthetic segments, *Canadian Journal of Linguistics*, 30: 415-451.

PRINCE, A. (1983) Relating to the grid, *Linguistic Inquiry*, 14: 19-100.

PUEL, M., NESPOULOUS, J-L., BONAFÉ, A., RASCOL, A. (1980) Etude neurolinguistique d'un cas d'anarthrie pure, *Grammatica*, 7: 239-292.

SELKIRK, E. (1982) The syllable, in van der Hulst H., Smith N. (Eds.) *The structure of phonological representation*, vol. 2., Dordrecht, Foris: 337-384.

SHATTUCK-HUFNAGEL, S., KLATT, D.H. (1979) The limited use of distinctive features and markedness in speech production: evidence from speech error data, *Journal of Verbal Learning and Verbal Behavior*, 18: 41-55.

SIEVERS, E. (1881) *Grundzüge der Phonetik*, Leipzig, Breitkopf, Härtel (5th edition, 1901).

STEMBERGER, J. (1984) Structural errors in normal and agrammatic speech, *Cognitive Neuropsychology*, 1: 281-313.

STEMBERGER, J., TREIMAN, R. (1986) The internal structure of consonantal cluster, *Journal of Memory and Language*, 25: 163-180.

TROUBETZKOY, N. (1970) *Principes de phonologie*, Paris, Klincksieck.

VALDOIS, S. (1990) Internal structure of two consonant clusters, in Nespoulous, J-L., Villiard, P. (Eds.) *Morphology, phonology and aphasia*, New York, Springer Verlag: 253-269.

WHITNEY, W. (1865) On the relation of vowels and consonants, *Journal of the American Oriental Society*, 8: 357-373.

M. VENTURA, PH. PAQUIER, J. DELEVAL

ACUTE NEUROGENIC SPEECHLESSNESS

INTRODUCTION

Acute loss of verbal expression after a neurological insult can prove to be challenging for the clinician. More than the neuropathological origin, the neurophysiological mechanisms which subtend the symptoms often constitute the major question to address. This is illustrated by the retrospective study of a patient we have seen in the acute phase of her illness, and have followed-up for $1\frac{1}{2}$ year.

CASE REPORT

Medical history

On April 29, 1993, a 72-year-old, French-speaking, right-handed woman was admitted to hospital following the sudden onset of paresis of the inferior part of the right hemiface and of the right upper limb, together with mutism. In the past she had suffered a left lower limb amputation after having remained for hours in a bombed railway carriage during World War II, and had been severely claustrophobic ever since. Later, she underwent right nephrectomy secondary to abscedation due to lithiasis, and cholecystectomy. In August 1991 she sustained a right posterior hemispheric stroke causing a residual left quadranopia. She was treated for hypertension with a calcium antagonist, and for asthma with anticholinergic puffs. She was a married housewife with a middle school education, and had two children. She was a nonsmoker, and did not take alcohol or illicit drugs.

Upon admission the blood pressure was 165/100 mm Hg, and the pulse was 92 and regular. Some sibilances were heard at chest auscultation. The remainder of the general physical examination was unremarkable. Neurological examination showed a conscious, collaborative, and adequate patient. Although she was unable to utter any vocal sound, auditory comprehension was normal according to a bedside screening. Spontaneous writing contained some grammatical and spelling errors. Cranial nerves examination disclosed a left superior quadranopia, a slight right-sided inferior facial weakness, a deviation of the tongue to the right, and an absence of gag reflex on the right. Except for the absent gag reflex, bedside testing of cranial nerves IX and X was strictly normal, though the patient complained in writing of minor swallowing difficulties with liquids. There was

Y. Lebrun (ed.), From the Brain to the Mouth, 81–97.

slight paresis of the entire right upper limb with impaired finger agility. Reflexes were slightly brisker on that side. Gait was not impaired, taking into account the fact that this well-compensated unijambist wore a lower limb prosthesis. Tests of sensation were normal, but coordination was slightly affected in the paretic limb. A CT scan of the brain disclosed only a right occipital hypodensity compatible with an old infarct. EEG revealed a left fronto-temporal slow wave dysrhythmia. One week after admission, a repeat CT scan revealed a left hemispheric ischaemic lesion in the territory of the left middle cerebral artery, in addition to the right occipital lesion. The new lesion involved the pars opercularis of the left frontal lobe (area 44) and the inferior part of the left premotor region (area 6) (Figure 1). An MRI of the brain had to be discontinued because the patient experienced a severe attack of claustrophobia during the procedure. She later refused any other attempt.

The neurological state of the patient evolved quite rapidly. One day after admission, there was complete resolution of the limb and facial paresis. The next day the patient was able to utter some undifferentiated gruntings, and 2 days later she succeeded in producing more elaborated raucous sounds. She no longer complained of swallowing difficulties.

The patient's neurological condition improved steadily in the following days, and it was possible to discharge her 3 weeks after admission. At that time she was capable of uttering a few short words quite effortfully. Grammatical errors in writing had disappeared, and she continued to use this communication channel as her favourite means of expression.

The patient received speech therapy for $1\frac{1}{2}$ year. She was seen again three times at our outpatient clinic. Except for the favourable evolution of the articulatory disorder, her physical and neurological condition did not change significantly during follow-up.

Language examination

The first formal assessment was performed 4 days after onset of the neurological symptoms, by means of routine bedside screening tests for aphasia which had been designed in our neurological department. Table 1 displays the patient's scores on the tests administered. There was no significant impairment of auditory comprehension, whereas oral expression consisted only of effortful productions of raucous sounds whenever the patient tried to speak. Reading aloud was not possible, but reading comprehension was adequate. Writing to dictation was normal at elementary level. Fifteen days post-onset, the patient's performance levels on a repeated bedside testing remained basically unchanged (Table 1).

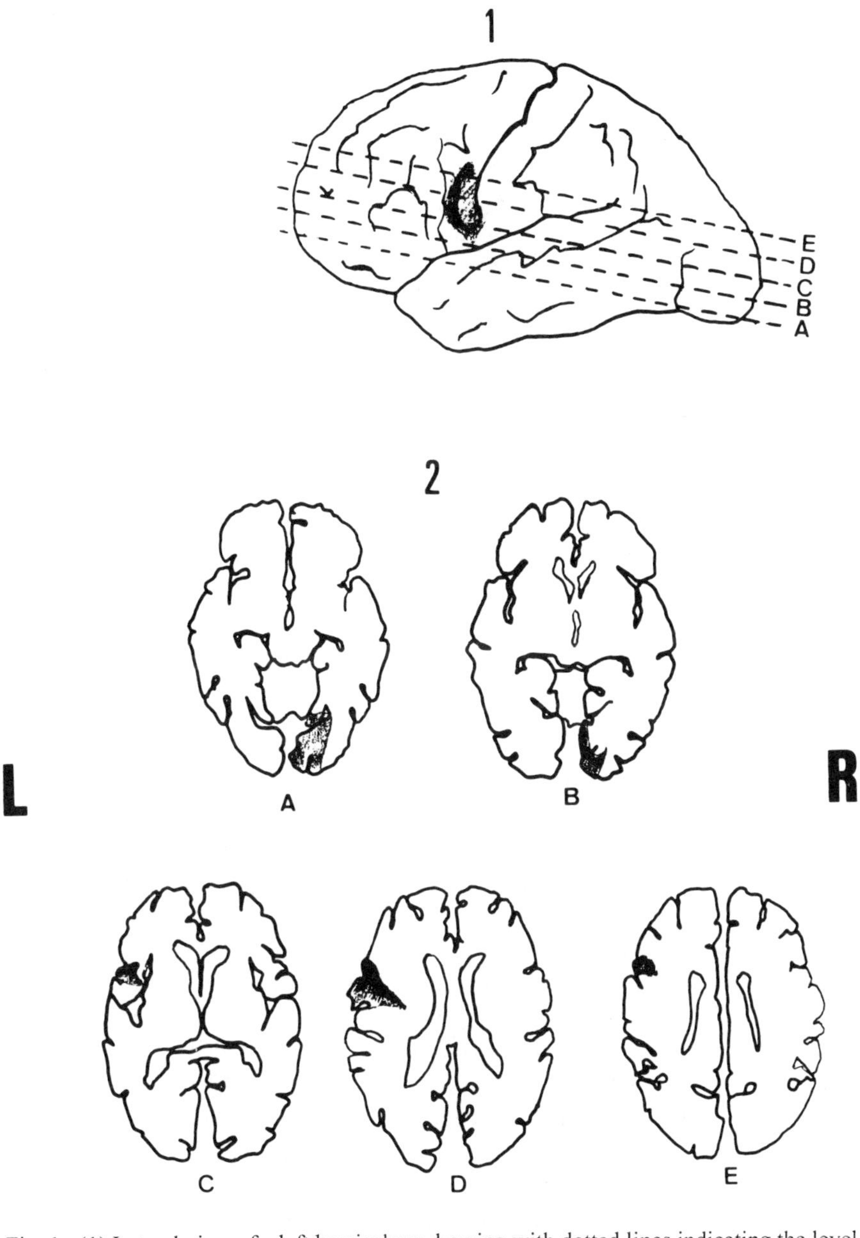

Fig. 1. (1) Lateral view of a left hemisphere drawing with dotted lines indicating the level and incidence of the template cuts (A→E) which correspond to the axial cuts of the CT scan performed one week post-onset. The cortical area of recent infarction encompasses the left frontal operculum and premotor region. (2) Templates with charted lesion. The right occipital lesion (A-B) corresponds to the old infarct. The area of recent infarction (C-D-E) involves the pars opercularis of the left frontal lobe (Brodmann's area 44) and the inferior part of the left premotor region (Brodmann's area 6), and extends subcortically into the subjacent white matter. L = left; R = right.

TABLE 1
Patient's peri-acute performance levels on routine bedside screening tests for aphasia, expressed as numbers of correct responses

Language Modality	Tests	Time post-onset 4 days	15 days
Verbal output	Phrase length	0/10	1/10
	Automatized sequences	0/10	0/10
	Repetition: phonemes	2/5	3/5
	words	0/10	1/10
	sentences	0/5	0/5
	Confrontation naming	0/10	0/10
Auditory comprehension	Object pointing task	10/10	10/10
	Commands	10/10	10/10
Written language	Reading aloud: letters	1/3	2/3
	words	0/5	0/5
	sentences	0/2	0/2
	Writing	5/5	5/5
Reading comprehension	Word/picture matching	10/10	10/10
	Commands	5/5	5/5

At that time, the French adaptation of the Boston Diagnostic Aphasia Examination (BDAE) (Goodglass and Kaplan, 1972; Mazaux and Orgogozo, 1983) was administered. Figure 2 shows the patient's profile which is consistent with the diagnosis of profound nonfluency. Auditory and reading comprehension were within normal limits. Writing was slightly dysorthographic (the patient was known to have experienced minor spelling difficulties premorbidly), and contained occasional omissions of words. Spontaneous writing, copying, dictation, and written confrontation naming were otherwise unremarkable (Figures 3 and 4). Graphomotricity was judged to be similar to the premorbid level (the patient was not a skilled writer). Verbal output was abolished, the patient being capable of only effortful grunts. An additional test for oral praxis (De Renzi et al., 1966) was performed without errors. However, mild praxic abnormalities were observed during the execution of rapidly alternating labio-lingual movements.

Two other language tests were also administered in the same period. The shortened version of the Token Test (De Renzi and Faglioni, 1978) could not disclose any significant auditory comprehension deficit (34/36 correct). The patient's responses on a written version of a visual confrontation naming test (Bachy-Langedock, 1989) were also within normal limits (35/36 correct).

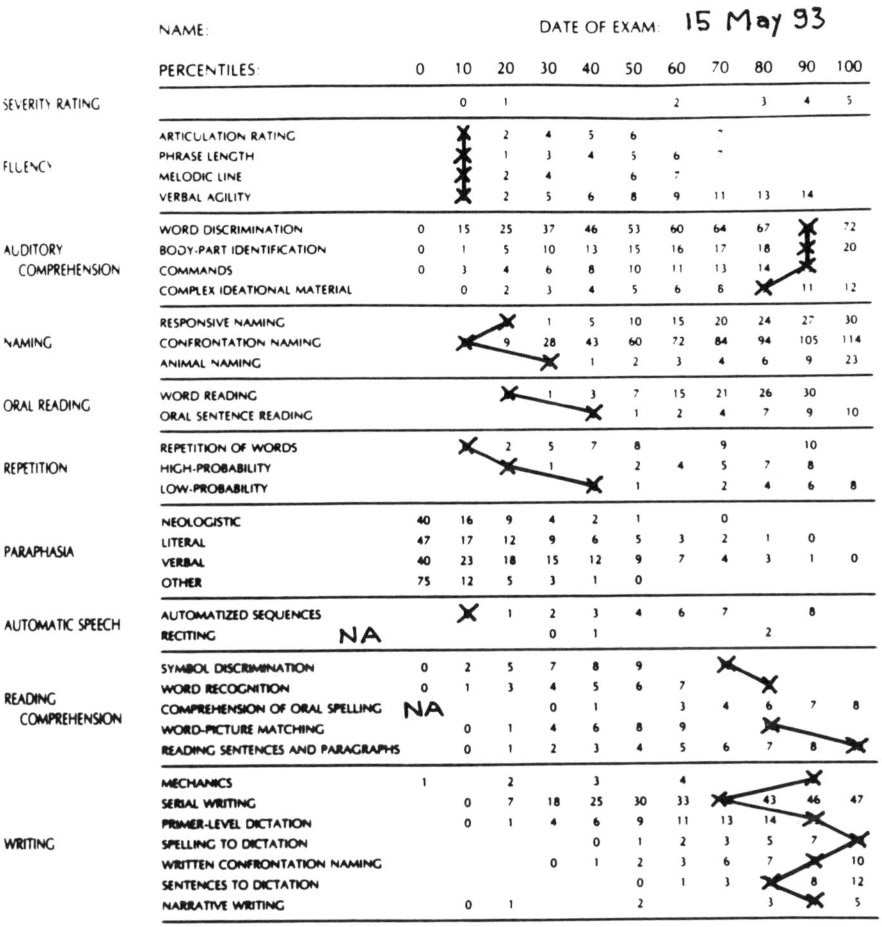

Fig. 2. Patient's BDAE subtest score profile. NA = not administered.

Language therapy was subsequently undertaken, and continued until December 1994. Gradual improvement occurred over a period of 6 months. By that time, verbal output had become rather efficient, but a severe articulatory disorder remained, speech being slow, dysprosodic, and syllabic. There were frequent substitutions, additions, omissions, and distortions of phonemes. Numerous hesitations were observed, as well as marked anomalies in scansion and rhythm. Syntax was normal. The patient did not experience word-finding difficulties. One year post-onset, an extended repetition task was presented in order to analyse the type of articulatory errors (see below). Repetition of isolated phonemes appeared to be normal.

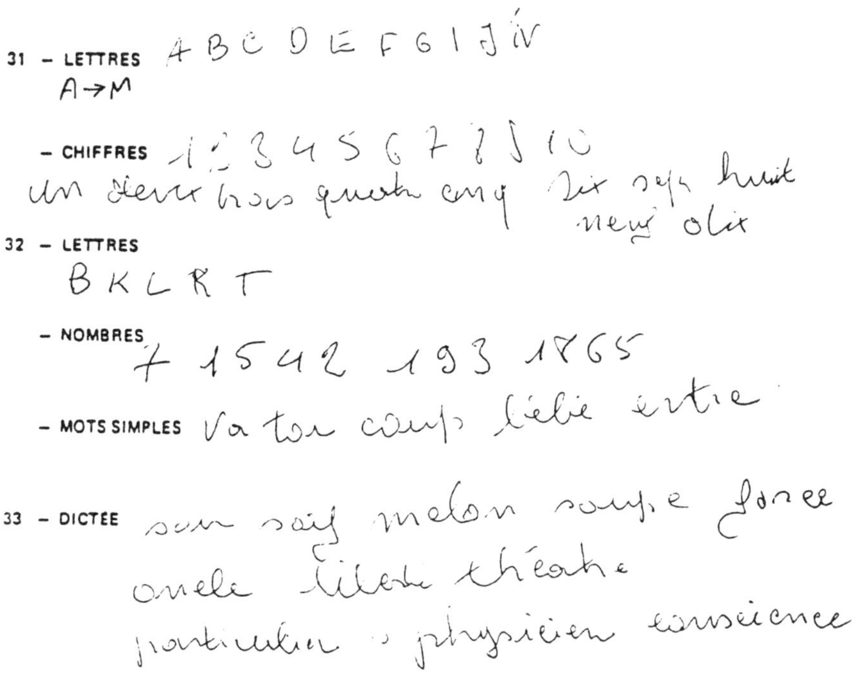

Fig. 3. Patient's performance on the BDAE subtests *Serial Writing*, *Primer Level Dictation*, and *Spelling to Dictation*. Scores are displayed on the BDAE summary profile (Figure 2).

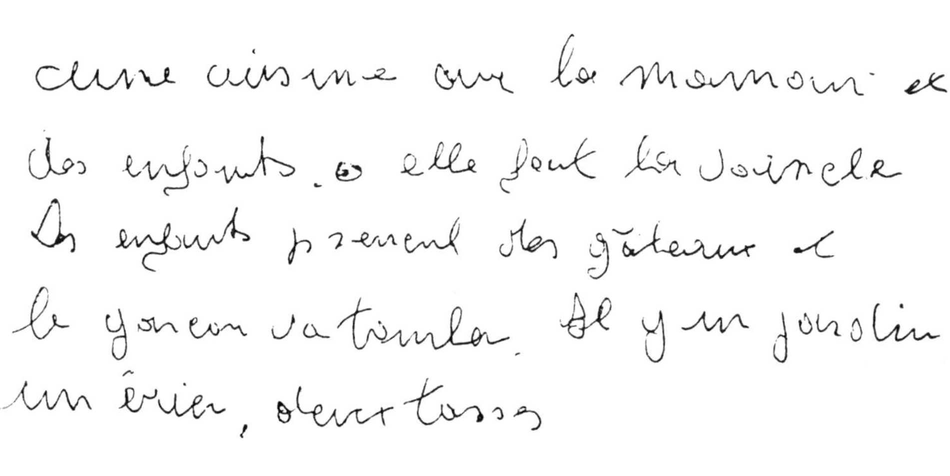

Fig. 4. Patient's written description of the "cookie theft" picture from the BDAE.

As to auditory and reading comprehension, the patient scored 34/36 on the Token Test, and 10/10 on the BDAE sentences and paragraphs. The written description of the Cookie Theft Picture from the BDAE was normal except for minor dysorthographic difficulties which were judged to be similar to the premorbid ones.

Neurolinguistic investigation

In order to determine the type of articulatory errors, one of us (JD) administered an extended repetition task to the patient. This task consisted of a first, experimental list of 204 isolated words, comprising 47 monosyllabic, 86 bisyllabic, 43 trisyllabic, and 28 tetrasyllabic words. The second list, published by Puel and co-workers (1981), contains 169 isolated words, namely 68 monosyllabic, 50 bisyllabic, 17 trisyllabic, and 34 tetrasyllabic stimuli, all chosen according to various phonetic and phonological criteria. The total number of phonemes was 789, namely 363 vowels and 426 consonants. The articulatory errors were analysed according to the method described by Puel and collaborators (1981). The stimuli were presented one by one, the patient being allowed to correct herself whenever she wanted to. Answers were tape-recorded, and transcribed using the International Phonetic Alphabet (IPA).

A preliminary analysis of the patient's responses on both lists revealed a variety of articulatory anomalies mainly at phonetic/phonological level, such as substitutions and transformations, omissions and reductions, additions, anticipations and metatheses, and displacements. Table 2 illustrates these anomalies and provides, in addition, an example of syllabication. A quantitative analysis of the responses on the second list shows 68 deviations (8.6%) out of a total of 789 phonemes. This corresponds to 64 (38%) words produced erroneously out of a total of 169 words.[1] Involvement of vowels was minimal (3% of deviations), whereas 13% of consonants were altered. Nineteen (28%) out of 67 consonantal clusters were simplified. Phonemic substitutions/transformations and omissions had a high occurrence, the former (62%) being more frequent than the latter (32%). Phonemic additions (epentheses) and anticipatory additions each accounted for 3% of the errors. Metatheses and displacements did not occur in the second list. Main types of substitutions/transformations were occlusivation of fricatives (24%), devoicing of voiced phonemes (16%), and errors on the articulation point (28%) with a tendency to produce bilabial phonemes.

Both the extended repetition task and the samples of conversational

[1]On the first list, 77 (38%) out of the 204 words were mispronounced.

TABLE 2

Overview of the patient's articulatory errors, and examples of mispronunciations made in the repetition tasks

Error type	Stimulus word [with English translation]	Target pronunciation	Patient's mispronunciation
1. Substitutions / transformations			
· aperturization	horizon [horizon]*	/ɔRizɔ̃/	/ɔRɛzɔ̃/
· voicing of voiceless phonemes	hélicoptère [helicopter]	/elicɔptɛR/	/elicɔbdɛR/
· devoicing of voiced phonemes	gros [thick]	/gRo/	/kRo/
· affricativation	flot [flow; flood]	/flo/	/pflo/
· occlusivation	frein [brake]	/fRɛ̃/	/pRɛ̃/
· delayed nasal resonance	armoire [cupboard]	/aRmwaR/	/aRbmwaR/
· nasalization	banc [bench]	/bɑ̃/	/mɑ̃/
· alveo-dentalization	jupon [petticoat]	ʒypɔ̃/	ʒytɔ̃/
2. Omissions / reductions			
· /R/ in consonantal clusters	structure [structure]	/stRyktyR/	/styətyR/
· /s/ in consonantal clusters	impérialiste [imperialist]	/ɛ̃peRjalist/	/ɛ̃peRjalit/
· /j/ before vowel	excursion [excursion]	/ɛkskyRsjɔ̃/	/ɛskyRsɔ̃/
· reductions of diphthong	nuit [night]*	/nwi/	/ni/
3. Additions			
· epentheses (schwa)	spectacle [spectacle]	/spɛktakl/	/səpɛktaklə/
· anticipatory	joue [cheek]	/ʒu/	/uʒu/
· others	avocat [lawyer]	/avɔka/	/avɔkwa/
4. Metatheses			
· "pure"	porte [door]*	/pɔRt/	/pɔtR/
· with omission	verglas [glazed frost]*	/vɛRgla/	/vɛlga/
5. Displacements			
· retrograde	plastique [plastic]*	/plastik/	/plakti/
· anterograde	imprudent [imprudent]*	/ɛ̃pRydɑ̃/	/ɛ̃pydRɑ̃/
6. Miscellaneous	perfectionner [to improve]*	/pɛRfɛksjɔne/	/pɛk - pɛk - pɛ - pɛ - fɛstɔne/
7. Syllabications	Tchécoslovaquie [Czechoslovakia]*	/tʃɛkɔslɔvaki/	/sɛk - kɔ - slɔ -vakijə/

* Examples taken from the experimental list of 204 words.

speech recorded during language therapy allow to summarize the patient's main speech characteristics as follows:

1. Vowel production is less disturbed than production of consonants.
2. Articulatory errors are unpredictable with regard to the moment of error production and the position within the word, but there exists a relative consistency as to the type of some errors (e.g. frequent occlusivation of fricative consonants).
3. Phonemes are substituted in that they often possess a faulty distinctive feature (e.g. devoicing of voiced phonemes).
4. The serial ordening of phonemes is disturbed.
5. There is an overall nasalization.
6. Speech rate is variable though mostly slowed down. Pauses are irregularly spaced, and frequently placed inadequately.
7. There are frequent intersyllabic interruptions which result in scanning speech, but verbal output is not always isochronic. Syllabication does not always respect syllabic boundaries.
8. In connected speech, words are not linked together.
9. There is an equalization of stressed and unstressed syllables.

DISCUSSION

Differential diagnosis

The patient's clinical picture in the peri-acute stage was characterized by total abolition of oral expression, whereas auditory and reading comprehension, and writing had remained normal or near-normal. When confronted with such a peri-acute picture, the clinician must consider different possibilities in order to reach differential diagnosis, since due to the generalized but transitory suppression of neural activity that occurs immediately after a sudden insult to the brain, the functioning of regions which are functionally or anatomically connected with the area of structural damage may be disturbed by deprivation of synaptic activity[2] (Kertesz, 1984). These different conditions may be:

[2] Von Monakow (1914) introduced the term *diaschisis* to denote this phenomenon.

Aphasia[3]: According to Darley and co-workers (1975), aphasia is a multimodal loss or reduction, due to brain damage, in the capacity to decode and encode meaningful linguistic elements. It is manifested in difficulties in listening, speaking, reading, and writing. In general, a remarkable dissociation between the total abolition of spoken output and the relative preservation of the other linguistic functions, including writing, may lead to focus attention on the former, and to neglect or minimalize possible anomalies in the latter. In the acute stage, however, speechless patients with good auditory comprehension capacity may show writing disturbances consisting in word-finding difficulties, paragraphic errors, perseverations, and ungrammatical productions (Schiff et al., 1983; Puel et al., 1984; Pellat et al., 1991; Kaminski et al., 1992). When in the course of recovery speechlessness eventually improves to lexically and syntactically near-normal level but writing remains disturbed, an aphasic disorder may be assumed (Kaminski et al., 1992). In those speechless patients in whom, in addition to the writing disorder, expressive aphasic anomalies become discernable as spoken output gradually returns, a diagnosis of Broca's aphasia may prove to be the correct one. In light of the grammatical and spelling errors in our patient's spontaneous writing in the acute stage, an aphasic deficit might be invoked to explain the nature of her communication disorder. The rapid resolution of the grammatical writing errors, and the rapidity with which the patient regained her premorbid writing level, however, contrast with the slow and laborious evolution of the longstanding articulatory disorder, thus suggesting that the former might be related to the wearing off effects of diaschisis, whereas the latter would rather depend on the structural lesion of the brain. Consequently, we may assume that the selective articulatory disorder which remained in our patient after the distant effects of the neural shock had worn off, is not aphasic in nature.

Dysarthria: According to Darley and co-workers (1975), dysarthria is "a collective name for a group of related speech disorders that are due to disturbances in musculature control of the speech mechanisms resulting from impairment of any of the basic motor processes involved in the execution of speech [p.2]". The nature of these neurogenic motor speech disorders is most frequently paralytic, atactic, dystonic, or myopathic. An extreme form of motor speech disorder is encountered in the so-called Foix-Chavany-Marie syndrome, i.e. the cortico-subcortical type of suprabulbar or pseudo-bulbar palsy which associates severe dysarthria and a bilateral central voluntary paresis of lower cranial nerves with preserved automatic,

[3]We do not consider here the articulatory disorders in the semiological context of Broca's aphasia, in which agrammatism, word-finding difficulties, and specific syntactic comprehension deficits are observable from the outset.

involuntary, emotional innervation (Weller, 1993). In the acute phase of a hemispheric insult, a sudden and total speechlessness without significant impairment of auditory and written comprehension can be associated with disturbances of the voluntary and reflex activity of the bucco-facial musculature (Jude and Trabaud, 1928; Alajouanine et al., 1959; Tonkonogy and Goodglass, 1981; Sugishita et al., 1987; Pellat et al., 1991; Weller, 1993; Chesser et al., 1994; De Smet, 1994). As Lebrun (1989) pointed out, these motor problems may confuse the clinical picture and render diagnosis difficult, as it may be problematic at that stage to differentiate speechlessness of non-dysarthric origin from pseudo-bulbar dysarthria. Consequently, it may be necessary to wait for these patients to recover some ability to verbalize before the distinction between dysarthria and other types of speechlessness can be made. However, symptoms such as increased cranial nerves reflexes, drooling, dysphagia, and lability of emotional expression should help the clinician differentiating between pseudo-bulbar dysarthria and other instances of speechlessness.

Although pseudo-bulbar palsy was not considered in our patient, neurological examination upon admission disclosed pathological cranial nerves functioning, such as a slight right inferior facial palsy, absence of gag reflex on the right, and a deviation of the tongue to the right. The patient was experiencing minor swallowing difficulties. These symptoms, however, were very mild and transitory; a few days later they had completely disappeared, leaving the patient with a severe verbal output disorder which could not be explained by the presence of disturbances in musculature control of the speech mechanisms. We therefore assume that the persistent articulatory disorder of the patient is not of dysarthric nature.

Aphemia: The term was originally coined by Broca (1861) to denote the selective disturbance of articulate speech which his now-famous patient Leborgne ("Tan") had acquired after a cerebral lesion. According to Broca (1861), speech is abolished because the aphemic patient has an amnesia specific to patterns of articulatory movements in the bucco-phonatory apparatus, while the speech musculature itself is not paralyzed. Other language modalities, such as auditory and reading comprehension, and writing remain intact. After many years of terminological confusion,[4] the term has nowadays recovered its original meaning. Endorsing the view of Schiff and collaborators (1983), we reserve the term *aphemia* for a disruption of a specific network in the motor system responsible for the organization of simultaneous and sequential motor actions into well-articulated speech, resulting in suppression of speech production, while writing, and auditory

[4]For a review, see Darley and co-workers (1975), and Lebrun (1989).

and reading comprehension remain remarkably preserved. At onset, the aphemic patient is totally speechless or capable of producing only minimal and effortful grunts. There may be aphasic deficits due to diaschisis or peri-lesional oedema, but as they clear up relatively quickly, the patient is left with a speechlessness which contrasts sharply with the normalcy of the other linguistic functions. In this phase, the patient typically prefers to use written language to communicate instead of trying to express himself vocally. Rare instances of a long-standing and total speechlessness have been observed (Ladame, 1902). However, most aphemic patients start verbalizing effortfully after some days, thus displaying a selective impairment of speech production. Speech then is slow, laborious, breathy, irregularly paced, occasionally explosive, with abnormal prosody and restricted melodic range (Benson, 1979; Schiff et al., 1983). This poorly articulated, hypophonic speech may result in distortions of phonemes. Word-finding and syntax, however, are normal. We suggest that, when the main symptom is distorted and unpredictable articulation with audible vocalization, aphemia fades into anarthria (or apraxia of speech; see below). Both conditions are indeed closely related to one another (Benson, 1979), and some authors consider *aphemia* and *anarthria* to be synonymous because of their symptomatological similarities (Pierre Marie in Klippel, 1908; Ladame and von Monakow, 1908; Sugishita et al., 1987; Pellat et al., 1991). Given the rapid evolution of our patient's neurological symptoms, and the characteristics of her language profile in the peri-acute stage, we feel that the term *aphemia* best fits her communication disorder at that time.

Anarthria: In contrast to Darley and co-workers (1975), we do not consider anarthria to be a total speechlessness due to severe loss of motor function of speech musculature (this we would call "global dysarthria"). Rather, we would like to equate anarthria with apraxia of speech, in the way Marie considered it to be a praxic disorder in 1908 (Ballet, 1908). In line with Darley and colleagues (1975), and Clarke and co-workers (1992) we define apraxia of speech — thus, anarthria in our terminological use — as an articulatory disorder resulting from impairment, due to brain damage, of the capacity to program the positioning of speech musculature for the volitional production of phonemes and the sequencing of muscle movements for the production of words, in the absence of significant damage to the motor or sensory pathways directly controlling the articulatory musculature. This definition closely resembles the one we already used to define aphemia in one of its possible later stages, at the moment verbal output reappears and articulation becomes intelligible. As a matter of fact,

Sugishita and collaborators (1987) consider aphemia to represent at least two types of disturbance: (a) an inability to articulate all except a limited set of phonemes, or (b) an ability to utter most phonemes but with many inconsistent articulatory errors such as substitutions, additions, omissions, and distortions (this is anarthria/apraxia of speech). In the latter, a residual dysprosody remains almost invariably. Together with the altered speech rate and the distorted articulation, it then often gives the impression of a foreign accent (Benson, 1979; Schiff et al., 1983). Taking into account the constellation of articulatory and prosodic alterations in combination with the presence of normal lexical and syntactic abilities in our patient's spoken and written output, and in the absence of auditory and reading comprehension difficulties, we propose that her verbal behaviour in the subsequent course of recovery resembles the one several authors have described in cases of anarthria/apraxia of speech (Johns and Darley, 1970; Lebrun et al., 1973; Lecours and Lhermitte, 1976; Itoh et al., 1979; Klich et al., 1979; Puel et al., 1981).

The anatomical areas involved in aphemia and in anarthria

Anatomical data in the literature on aphemia are derived from both neuropathological and neuroradiological observations. However, due to the terminological confusion prevailing in the literature, and the similar symptomatological descriptions given of both aphemia and anarthria, it is not always possible to determine whether the cases reported as instances of aphemia do indeed correspond to the entity we have labeled as such. Reviewing the anatomical data might consequently induce inconsistencies, and the reader should keep this possibility in mind.

One of the first detailed descriptions of post-mortem examination in a case of "pure aphemia" comes from Ladame and von Monakow (1908). They found a large, haemorrhagic cystic lesion destroying the posterior third of the third frontal gyrus (foot of F3) and the inferior half of the precentral gyrus, sparing a small opercular slip that partially covered the insula of the left hemisphere. The frontal operculum, the second frontal gyrus (F2), as well as the median part of the postcentral and supramarginal gyri were undermined by the cyst, immediately beneath the cortex. Recent neuroradiological data are in line with this anatomopathological observation. Thus, Ruff and Arbit (1981) documented the case of a 15-year-old right-handed girl who presented with aphemia after drainage of a haematoma extending from the left posterior premotor cortex to the inferior precentral gyrus. A repeat CT scan 3 months after the surgical intervention showed residual damage extending from Broca's area to the left frontal operculum and inferior precentral gyrus. The authors concluded that these neuroradiological data were compatible with reported autopsy findings. In two of Schiff and co-workers' (1983) aphemic patients, a CT scan disclosed a lesion in the lower half of the left precentral gyrus, with subjacent white matter involvement. In the remaining two patients, the lesion involved the left anterior limb of the internal capsule between the head of the caudate nucleus and the putamen. Sugishita and colleagues (1987) reported on a 58-year-old right-handed man with aphemia following an infarct. Both CT scan and MRI showed that the lesion affected the lower parts of the left precentral and postcentral gyri and their underlying white matter, such as the middle region of the corona radiata and the left arcuate fasciculus. In Pellat and collaborators' (1991) case of traumatic aphemia in a 45-year-old woman, an MRI performed 19 months post-onset showed a residual porencephalic cavitation encompassing the foot of F3, the inferior part of the precentral gyrus, and the underlying subcortical regions.

Not only has structural damage to the opercular region been documented by CT scan and MRI, but functional dysfunctions have also been reported using PET imaging in cases where conventional CT scans had failed to disclose any corresponding structural lesion. Kushner and co-workers (1987) described a 74-year-old right-handed man with aphemia following cerebral infarction. While a CT scan without and with contrast showed no abnormality in the left hemisphere, a PET scan revealed a focus of hypometabolism near the inferior aspect of the precentral gyrus. Broca's area was not involved, but the area of hypometabolism extended to subjacent gray and white matter structures.

Concerning cases labeled as anarthria/apraxia of speech, both neuropathological and neuroradiological data point to clear anatomical similarities with the observations documented in cases of aphemia. Lesions involve mostly the inferior half of the left precentral gyrus, sometimes including the Rolandic operculum, sometimes extending towards the foot of F2 but not towards that of F3 (Lecours and Lhermitte, 1979; Itoh et al., 1979; Tonkonogy and Goodglass, 1981; Puel et al., 1984; Clarke et al., 1992).

Consequently, both neuropathological and neuroradiological data suggest that, in right-handers, the cerebral substrate for aphemia is damage to the inferior part of the left precentral gyrus, often with extension towards the foot of F3, while for anarthria/apraxia of speech it is damage to the inferior part of the left precentral gyrus, sometimes with extension towards the foot of F2. In our patient, the localization of the lesion, as disclosed by CT scan, fits in well with this finding (Figure 1).

Schiff and colleagues (1983) proposed that aphemia may result from a lesion in any one of three related areas where motor systems for speech intersect with brain mechanisms for language: (a) a small lesion of the pars opercularis, (b) a lesion limited to the lower half of the primary motor cortex, and (c) lesions underlying either of these areas from the immediately subcortical white matter to the anterior limb of the internal capsule. Mohr and co-workers (1978) speculated that, due to some synergistic interaction of the entire opercular and insular region, the subareas of the opercular region share common language functions, and partially overlap one another in contributing to language functions. Ojemann and Whitaker (1978) suggested that the frontal opercular area served as the intersection between the cerebral system for language and that for motor speech activity. Alexander and collaborators (1990) considered the opercular region to contain critical parts of different language systems. The close connection between neighbouring cortical and subcortical areas concerned with speech articulation, phonological programming, and phonological encoding may, therefore, explain some aspects of the similarities and overlaps between aphemia, anarthria, aphasia, and dysarthria.

CONCLUSION

Based upon neurological, neurolinguistic, and anatomical findings, it seems reasonable to conclude that, at onset of the neurological disorder, the patient we describe in the present study suffered from aphemia which subsequently evolved towards anarthria/apraxia of speech. This case illustrates the diagnostic difficulties the clinician encounters in the presence of acute neurogenic speechlessness, and shows that it is not always possible to use a distinct diagnostic label to define the real nature of such a sudden clinical picture. The clinician should keep in mind that sometimes it is necessary to wait until the patient has recovered some ability to verbalize before the type of acute speechlessness can be determined. In cases of acute aphemia, for instance, we suggest that four different clinical entities may possibly become clear in the post-acute stage, depending on the location of the lesion: (a) lasting aphemia, (b) anarthria/apraxia of speech, (c) dysarthria, and (d) aphasia. However, as the areas of brain responsible for these clinical entities lie close together, damage to a complex though relatively narrow neuro-axonal network, such as the one proposed by Schiff and co-workers (1983), may lead to semiological pictures which overlap one another. It appears, then, that to use the term *aphemia* to denote a condition of acute neurogenic speechlessness in which the willingness to communicate is preserved, does not bear any diagnostic implications regarding the subsequent evolution of the verbal output disorder.

REFERENCES

ALAJOUANINE, T., LHERMITTE, F., CAMBIER, J., RONDOT, P., and LEFEBVRE, J.P. (1959) Perturbations dissociées de la motricité facio-bucco-pharyngée avec aphémie dans un ramollissement sylvien profond partiel. *Revue Neurologique*, 100, 493-498.

ALEXANDER, M.P., NAESER, M.A., and PALUMBO, C. (1990) Broca's area aphasias: aphasia after lesions including the frontal operculum. *Neurology*, 40, 353-362.

BACHY-LANGEDOCK, N. (1989) *Batterie d'Examen des Troubles en Dénomination*. Bruxelles: Editest.

BALLET, G. (1908) Apraxie faciale (impossibilité de souffler) associée à de l'aphasie complexe (aphasie motrice et aphasie sensorielle); apraxie et aphémie. *Revue Neurologique*, 16, 445-447.

BENSON, D.F. (1979) *Aphasia, Alexia, and Agraphia*. New York: Churchill Livingstone.

BROCA, P. (1861) Remarques sur le siège de la faculté du langage articulé, suivies d'une observation d'aphémie (perte de la parole). *Bulletin de la Société d'Anatomie*, 6, 330-357.

CHESSER, M.Z., WILLIAMS, L.H., and ANGTUACO, E.J.C. (1994) Acute pseudobulbar mutism with good functional recovery. *Neurology*, 44, 176-177.

CLARKE, C.E., HOLDER, D.S., CURRIE, S., and STONE, J. (1992) Cerebral localisation in articulatory dyspraxia. *Journal of Neurology, Neurosurgery, and Psychiatry*, 55, 168-169.

DARLEY, F.L., ARONSON, A.E., and BROWN, J.R. (1975) *Motor Speech Disorders*. Philadelphia: W.B. Saunders Company.

DE SMET, Y. (1994) The "anterior operculum syndrome". *Journal of Neurology*, 241, 570-571.

DE RENZI, E., and FAGLIONI, P. (1978) Normative data and screening power of a shortened version of the Token Test. *Cortex*, 14, 41-49.

DE RENZI, E., PIECZURO, A., and VIGNOLO, L.A. (1966) Oral apraxia and aphasia. *Cortex*, 2, 50-73.

GOODGLASS, H., and KAPLAN, E. (1972) *The Assessment of Aphasia and Related Disorders*. Philadelphia: Lea & Febiger.

ITOH, M., SASANUMA, S., and USHIJIMA T. (1979) Velar movements during speech in a patient with apraxia of speech. *Brain and Language*, 7, 227-239.

JOHNS, D.F., and DARLEY, F.L. (1970) Phonemic variability in apraxia of speech. *Journal of Speech and Hearing Research*, 13, 556-583.

JUDE, and TRABAUD (1928) Hémiplégie gauche avec anarthrie; accès de fou rire contrastant avec la correction de la mimique douloureuse. *Revue Neurologique*, 49, 726-728.

KAMINSKI, H.J., ADAMS, N., BURNSTINE, T.H., CIVIL, R.H., and RUFF, R.L. (1992) Relation of aphemia and agraphia. *European Neurology*, 32, 302-304.

KERTESZ, A. (1984) Recovery from aphasia. In F.C. Rose (ed) *Progress in Aphasiology. Advances in Neurology* (Vol. 42). New York: Raven Press; 23-39.

KLICH, R.J., IRELAND, J.V., and WEIDNER, W.E. (1979) Articulatory and phonological aspects of consonant substitutions in apraxia of speech. *Cortex*, 15, 451-470.

KLIPPEL, M. (1908) Discussion sur l'aphasie. *Revue Neurologique*, 16, 611-636.

KUSHNER, M., REIVICH, M., ALAVI, A., GREENBERG, J., STERN, M. and DANN, R. (1987) Regional cerebral glucose metabolism in aphemia: a case report. *Brain and Language*, 31, 201-214.

LADAME, P.L. (1902) La question de l'aphasie motrice souscorticale. *Revue Neurologique*, 10, 13-18.

LADAME, P.L., and VON MONAKOW, C. (1908) Observation d'aphémie pure (anarthrie corticale). *Encéphale*, 3, 193-228.
LEBRUN, Y. (1989) Apraxia of speech: the history of a concept. In P. Square-Storer (ed) *Acquired Apraxia of Speech in Aphasic Adults*. London: Taylor & Francis; 3-19.
LEBRUN, Y., BUYSSENS, E., and HENNEAUX, J. (1973) Phonetic aspects of anarthria. *Cortex*, 9, 126-135.
LECOURS, A.R., and LHERMITTE, F. (1976) The "pure form" of the phonetic disintegration syndrome (pure anarthria); anatomoclinical report of a historical case. *Brain and Language*, 3, 88-113.
MAZAUX, J.M., and ORGOGOZO, J.M. (1983) *Echelle d'Evaluation de l'Aphasie*. Issy-les-Moulineaux: EAP.
MOHR, J.P., PESSIN, M.S., FINKELSTEIN, S., FUNKENSTEIN, H.H., DUNCAN, G.W., and DAVIS, K.R. (1978) Broca aphasia: pathologic and clinical. *Neurology*, 28, 311-324.
OJEMANN, G.A., and WHITAKER, H.A. (1978) Language localization and variability. *Brain and Language*, 6, 239-260.
PELLAT, J., GENTIL, M., LYARD, G., VILA, A., TAREL, V., MOREAU, O., and BENABID, A.L. (1991) Aphemia after a penetrating brain wound: a case study. *Brain and Language*, 40, 459-470.
PUEL, M., NESPOULOUS, J.L., BONAFE, A., and RASCOL, A. (1981) Etude neurolinguistique d'un cas d'anarthrie pure. *Grammatica VII*, 1, 239-291.
PUEL, M., NESPOULOUS, J.L., CARDEBAT, D., LECOURS, A.R., and RASCOL, A. (1984) Localization of lesions in pure anarthria. In D. Caplan, A.R. Lecours, and A. Smith (eds) *Biological Perspectives on Language*. Cambridge: MIT Press; 255-263.
RUFF, R.L., and ARBIT, E. (1981) Aphemia resulting from a left frontal hematoma. *Neurology*, 31, 353-356.
SCHIFF, H.B., ALEXANDER, M.P., NAESER, M.A., and GALABURDA, A.M. (1983) Aphemia: clinical-anatomic correlations. *Archives of Neurology*, 40, 720-727.
SUGISHITA, M., KONNO, K., KABE, S., YUNOKI, K., TOGASHI, O. and KAWAMURA, M. (1987) Electropalatographic analysis of apraxia of speech in a left hander and in a right hander. *Brain*, 110, 1393-1417.
TONKONOGY, J. and GOODGLASS, H. (1981) Language function, foot of the third frontal gyrus, and Rolandic operculum. *Archives of Neurology*, 38, 486-490.
VON MONAKOW, C. (1914) *Die Lokalization im Grosshirn und der Abbau der Funktion durch kortikale Herde*. Wiesbaden: Bergmann.
WELLER, M. (1993) Anterior opercular cortex lesions cause dissociated lower cranial nerve palsies and anarthria but no aphasia: Foix-Chavany-Marie syndrome and "automatic voluntary dissociation" revisited. *Journal of Neurology*, 240, 199-208.

ACKNOWLEDGMENTS

The authors acknowledge the patient's willingness to provide the radiological and audiotaped documents which allowed this study to be worked out. Mrs. Anne Leonard (Dept. of Neurology, Hôpital Universitaire Erasme U.L.B.) assessed the patient in the periacute stage.

H. VAN DONGEN, P. VAN DOORN, E. YOUSEF-BAK, J. POOT

SLOWLY PROGRESSIVE DYSARTHRIA

INTRODUCTION

Dysarthria, a motor speech disorder resulting from a disturbance of the neuromuscular control of the speech organs, is a relatively frequent symptom in a number of neurological affections. In the Dutch transient ischemic attack trial study (Koudstaal et al., 1993) 24% of the patients having suffered a TIA or a nondisabling stroke had dysarthria. The incidence of dysarthria was superior to that of aphasia (19%). In Dippel's investigation (1994) of subarachnoid hemorrhage, dysarthria was diagnosed on 6% of the cases. In a consecutive series of 17 children with posterior fossa tumours, dysarthria was observed 7 times, i.e. 41% (van Dongen et al., 1994). Like acquired stuttering (Lebrun et al., 1983), dysarthria may be the first presenting symptom of a neurological disease (Oosterhuis, 1984). Lamprecht (1990) found that dysarthria, dysphagia and/or dysphonia were the initial symptom(s) in about 30% of patients with myasthenia gravis or amyotrophic lateral sclerosis.

Following a stroke, dysarthria may be the only observable symptom. This pure or isolated dysarthria is usually mild and transient (Ichikawa and Kageyama, 1991). At times, transient dysarthria due to stroke is accompanied by facial paresis (Kim, 1994).

CASE REPORT

In July 1992, a 72-year-old, righthanded woman came to the outpatient clinic at the University Hospital in Rotterdam complaining of persistent speech difficulties which had begun 4 years earlier. She stated that the difficulties had had a sudden onset, following the unexpected death of her husband.

Some time after the beginning of her dysarthria she was given speech therapy, but to no avail. In 1990, her GP referred her to a neurologist. However, apart from her dysarthria, no neurological deficit could be detected. The EEG and the CT-scan were not contributive. Accordingly, the possibility that her speech disorder might be psychogenic was considered. Maybe it had originated as a reaction to the painful loss which the patient experienced when her husband died.

Y. Lebrun (ed.), From the Brain to the Mouth, 99–102.

When the patient was examined at the university hospital, it was found that she had mild dysarthria with imprecise consonant articulation. Deliberate nonverbal lingual movements were somewhat clumsy and tongue-clacking was impossible. However, there was no dysphagia.

Language examination disclosed no aphasic deficit and the patient's score on the Progressive Matrices was normal. Memory was not impaired. The patient appeared to lead an active life including car-driving and daily visits to the swimming-pool. She lived on her own. She felt fit. However, she had been on medication for the last ten years to control hypertension. CT-scan was again not contributive and the possibility that the dysarthria had a psychological origin was again considered. Repeat examination took place three months later. Apart from the dysarthria no neurological impairment could be detected. Nonverbal lingual movements now appeared normal. Laryngoscopic examination of the vocal folds disclosed no anomalies.

The patient was accompanied by her daughter-in-law who stated that the speech difficulties developed *gradually* a year or so before the death of the patient's husband. Initially, the dysarthria was very light but it had worsened with the passing of time. Despite the information provided by the daughter-in-law, the idea that the dysarthria might be functional was again entertained. Speech therapy was recommended. The patient was seen again in April 1994. It appeared that, despite speech therapy, dysarthria had worsened. The voice sounded harsh and strained-strangled. There was moderate hypernasality, slow delivery, and audible inspiration. Some consonants were distorted, others were omitted, particularly at the end of words. Neurological examination revealed mild deficits: a mild diplegia facialis and mild disturbances of the tongue movements. Magnetic resonance imaging showed small anomalies in the cerebral white matter and the basal ganglia, particularly the putamen, bilaterally. A *vessel disease* was assumed, probably caused by the long standing hypertension (cp. Coffens and Robey in this volume).

The deterioration of speech continued after this examination. In September 1994 oral expression was almost impossible. Only monosyllabic words could be sporadically formed. The voice was strained-strangled, and occasionally harsh. The patient's dysarthria was considered to be of the pseudobulbar type. Because of her severe motor speech disorder the patient had to communicate in writing. She did so without making spelling errors or showing word-finding difficulties. Comprehension of spoken language was unimpaired.

COMMENT

In this case, then, dysarthria was for more than two years the only complaint and only observable deficit Electroencephalography and neuroradiology failed to reveal any organic damage that might account for the motor speech disorder, so that a psychogenic etiology was repeatedly considered. Yet, whilst psychological problems and psychic stress may certainly affect voice production (hysterical mutism), it is doubtful whether they can cause dysarthria. Dysarthria is, in essence, an organic disturbance. If it has a sudden onset and disappears fairly rapidly, a transient ischemic attack or mild cerebro-vascular accident may be assumed. If it develops slowly and worsens over time, some degenerative or space-occupying process should be suspected.

As the present case shows, the patient may not become immediately aware of his nascent dysarthria. A certain time may elapse before he realises that his motor speech is affected. And he may then be tempted to ascribe the disturbance to some painful psychological experience.

When dysarthria comes about gradually and for an appreciable period of time remains the patient's sole deficit, it reminds one of primary progressive aphasia. The latter is a slowly progressive aphasia that develops in the absence of other significant cognitive impairments. Because its beginning is insidious, it may pass unnoticed for some time. Moreover, in patients with recognised primary progressive aphasia, the EEG, the CT-scan and the MRI may be normal initially. However, PET or SPECT scan will generally reveal hypofusion in the left hemisphere (Duffy and Petersen, 1992). Accordingly, such scans may also be indicated in cases with primary progressive dysarthria, when EEG, CT-scan and MRI are not contributive.

Primary progressive aphasia may eventually become total or global, just as primary progressive dysarthria may evolve into speechlessness or quasi-speechlessness, as in the present case.

It appears, then, that persistent isolated dysarthria should not be offhandedly discharged as functional, but should be carefully followed up, as it is indicative of a progressive pathological process affecting the brain.

REFERENCES

Dippel D. (1994) *Decision analysis in the clinical neurosciences*. Amsterdam, Thesis Publishers.

Duffy J, Petersen R. (1992) Primary progressive aphasia. *Aphasiology* 6:1-15.

Ichikawa K, Kageyama Y. (1991) Clinical anatomic study of pure dysarthria. *Stroke* 22:809-812.

Kim J. (1994) Pure dysarthria, isolated facial paresis, or dysarthria-facial paresis syndrome. *Stroke* 10:1994-1998.

Koudstaal P, Algra A, Van Gijn J, Kapelle J, Pop G, Van Latum J, Van Sieten J. (1993) The Dutch TIA Trial Study Group. Predictors of major vascular events in patients with a transient ischemic attack or nondisabling stroke. *Stroke* 24:527-531.

Lamprecht A. (1990) Dysarthrie, Dysphagie oder Dyspnoe als Grund der Erstvorstellung bei Myasthenia gravis pseudoparalytica und bei amytropher Lateralsklerose. *Laryngorhinootologie* 69:48-51.

Lebrun Y, Rétif J, Kaiser G. (1983) Acquired stuttering as a forerunner of motor-neuron disease. *Journal of Fluency Disorders* 8:161-167.

Van Dongen H, Catsman-Berrevoets C, Van Mourik M. (1994) The syndrome of 'cerebellar' mutism and subsequent dysarthria. *Neurology* 44:2040-2046.

DYSFLUENCY

Y. LEBRUN

ADULT-ONSET STUTTERING

INTRODUCTION

Adult-onset stuttering

The label *adult-onset stuttering* refers to a stutter which is observed in an adult who never stuttered before. Typically, the individual's speech develops normally in childhood, and there is no dysfluency in infancy or adolescence. At some point in adult age, however, verbal delivery becomes suddenly or progressively halting, with involuntary blocks, repetitions and/or prolongations, possibly accompanied by unintentional muscle contractions primarily involving the face. *Adult-onset stuttering*, then, denotes an acquired speech impediment occurring in a grown-up who always spoke fluently in the past.

This definition implies that from a chronological point of view adult-onset stuttering is different both from developmental and from recurring stuttering. *Developmental stuttering* is a speech impediment which develops in childhood. It may or may not subside as the individual grows older. If it does not disappear with age, it tends to become a lifelong condition, unless therapy succeeds in eliminating it. As for *recurring stuttering*, it is observed in people who had developmental stuttering in childhood, eventually got over it and then in adulthood start to stutter again. Recurring stuttering, then, is a relapse at adult age of remitted developmental stuttering (see Van Borsel and Cappaert in this volume).

Adult-onset stuttering occurs in grown-ups who have always shown age-appropriate speech delivery skills. It involves a reduction or limitation of these skills. It is a loss. The patient speaks now less fluently than he used to. Adult-onset stuttering is therefore considered to be an *acquired* speech disorder.

Neurogenic stuttering

Frequently adult-onset stuttering occurs in conjunction with brain damage. Shortly after having sustained a cerebral injury or in the course of an encephalitic disease the patient becomes dysfluent. In such a case, a causal relationship is generally assumed between the neurological condition and the stutter. The latter is accordingly called *neurogenic* or *neurological stuttering*. When the causal lesion is limited to the cortex, and possibly

Y. Lebrun (ed.), From the Brain to the Mouth, 105–138.

the immediately subjacent white matter, some clinicians speak of *cortical stuttering.*

The existence of neurogenic stuttering has long been recognized. As early as 1817, Jean-Marc Itard, in his *Mémoire sur le Bégaiement* published in the *Journal Universel des Sciences Médicales*, noted that in adults stuttering could be consequent upon a stroke or upon adynamic fever.

Objections to the label "adult-onset stuttering"

Some clinicians strongly object to the use of the diagnostic label *adult-onset stuttering*. They find that the term *stuttering* should be reserved for the developmental condition. Only stuttering that begins in childhood deserves to be called stuttering. Dysfluency or a similar word should be used to refer to the acquired disorder which superficially resembles stuttering but is in fact different from it.

This claim rests on the view that developmental stuttering differs in nature from adult-onset stuttering, especially when the latter is neurogenic (Culatta and Leeper, 1988).

Developmental versus adult-onset stuttering

Various reasons have been put forward to support the idea that developmental and acquired stuttering are dissimilar in kind.

1. It has been claimed that the loci of dysfluency are not identical in the two conditions. Specifically, it is said that final sound repetitions are frequent in neurogenic stuttering, while they are conspicuously absent in developmental stuttering.

This contention, however, is not borne out by clinical facts. While it is certainly true that a number of patients with adult-onset stuttering make final sound repetitions, it is not true that developmental stutterers never do. On the contrary, it has been observed that children with developmental stuttering may very well iterate final sounds (Lebrun and Van Borsel, 1990) and that mentally handicapped adults with developmental stuttering may produce word-final repetitions and prolongations (Stansfield, 1995)

On the other hand, some patients with neurogenic stuttering never make final repetitions, as the cases reported by Rosenbek et al. (1978) show. Therefore, the loci of repetition do not clearly differentiate child-onset from adult-onset stuttering. The absence or presence of final sound repetitions or prolongations does not enable one to confidentially tell the developmental from the acquired condition.

2. It is often maintained that the repetitions do not involve the same

speech segments in developmental as in neurogenic stuttering. It is contended that in the former disorder iterations affect primarily sounds, part-words and short words, whereas in the latter there may be striking repetitions of phrases and even of sentences.

This claim does not accord well with clinical reality. In fact, there are patients suffering from neurogenic stuttering who repeat primarily, if not exclusively, sounds, syllables and short words (Lebrun and Leleux, 1985). Conversely, some developmental stutterers make phrase repetitions, as the following (translated) excerpt from a tape-recording of a French-speaking subject with child-onset stuttering demonstrates (Hyphens indicate clonic, and commas non-clonic repetitions):

"(Last summer) I went to N-N-N-N-N-N-Nor, I went to N-Normandy with, I went to Normandy with er my parents. As, as er, as er, as the weather was rather f, er as the weather was rather f-fine, that is, er we, we, we s, we stayed, we stayed in N-Normandy al, altho, altho, although we had first intended, although we had first intended er t-t-to go further s, to go further, to go further south, to go further south and to go further south and visit Brittany"

As this example shows, there may be striking phrase repetitions in the verbal output of developmental stutterers. Consequently, the speech units involved in the repetitions offer no safe criterion by which to distinguish developmental from adult-onset stuttering.

3. It has been argued that function words are more frequently stuttered upon in neurogenic than in developmental stuttering. The above excerpt shows, however, that conjunctions like *as* or *although*, which are functors, may very well be stumbled upon by developmental stutterers. Indeed, Bloodstein and Gantwerk (1967) have found that in young stutterers there is "a tendency for stuttering to occur unusually often on pronouns and conjunctions." In toddlers, repetition of the conjunction *and* or of the adverb *then* or of both may be particularly conspicuous. On the other hand, Baratz and Mesulam's patient (1981) who had acquired stuttering due to brain trauma stuttered only on content words. It follows that the parts of speech which are the most frequently stuttered upon do not clearly differentiate developmental from acquired stuttering.

4. According to some researchers, in developmental stuttering dysfluency increases with the spontaneity of the verbal output, conversation being more difficult than imitative speech or oral reading. In neurogenic stuttering, the reverse is found.

This view too fails to be verified by clinical observation. In fact, there are developmental stutterers who find imitative speech or oral reading as difficult as verbal intercourse. Conversely, there are patients with acquired stuttering who stutter less when reading aloud or repeating someone's words than when carrying on a conversation. The patient of Baratz and Mesulam (1981) who had neurological stuttering did not stutter while reading aloud but did stutter in conversation. Mazzuchi et al. (1981) noted that in their cases of neurogenic stuttering imitative speech and oral reading led to decreased dysfluency. A patient of Helm-Estabrooks (1986) stuttered in conversation and in naming and reading aloud tasks, but not when producing serial or imitative speech. As for the neurological patients of Horner and Massey (1983), Lebrun and Leleux (1985) and Lebrun et al. (1986), they evidenced stuttering in speaking but not in singing.

It follows that variation in the degree of dysfluency as a function of the propositionality or spontaneity of the verbal output is not a reliable touchstone for determining whether a stutter is developmental or acquired.

5. It has been stated that neurological stuttering is at times monosymptomatic, whereas developmental stuttering never is. Clinical findings, however, falsify the second part of this proposition. A number of toddlers with developmental stuttering have been found to produce only repetitions. At first, the little patient of Manders and Bastijns (1989) showed only word repetitions. And in the cases of developmental stuttering reported by Rudmin (1984) and by Mowrer (1987), final repetitions were the sole symptom in evidence.

It appears, then, that acquired and developmental stuttering can both be monosymptomatic. Therefore, the presence of only one symptom does not enable clinicians to differentiate between the two conditions.

6. Some authors assert that in developmental stuttering there is (often) an adaptation effect, whilst in acquired stuttering there is none. This assertion too fails to be borne out by clinical facts. Patients with adult-onset stuttering may speak more and more fluently as they read the same passage aloud over and over again. In a case of neurological stuttering observed by Quinn and Andrew (1977) there was a clear adaptation effect. In a patient who stuttered following brain injury, Rosenfield (1972) noted "a decreased frequency of facial contortions with repeated readings of the same passage." And in Koller's first case of stuttering resulting from an extrapyramidal disease (1983) "adaptation occurred immediately on repetitive effort."

Therefore, sensitivity of the dysfluency to such laboratory manipulations as repeated reading of the same material does not form an infallible criterion by which to distinguish between the two types of stuttering.

7. It has been submitted that patients with acquired stuttering do not develop so-called secondary features, that is to say, that they never present

with such concomitant involuntary movements as eye-blinking, grimacing or flaring of the nostrils. This proposition also runs counter to clinical reality. Patients with adult-onset stuttering may very well show involuntary muscle contractions during blocks or repetitions. For instance, in a neurological patient Rosenfield (1972) observed occasional facial contortions when she was trying to "get the word out." Grimacing during dysfluencies "was a salient feature" in Baratz and Mesulam's case of neurogenic stuttering (1981). And in the case of acquired stuttering reported by Lebrun et al. (1990) "blocks were often accompanied by involuntary blinking or closing of the eyes and by a slight bending of the head." The presence of so-called secondary features, then, does not differentiate developmental from acquired stuttering.

8. It has been asserted that patients with acquired stuttering are much less concerned about their speech disorder than patients with developmental stuttering. Clinical observations disprove this view, however. Rosenfield's neurological patient (1972) evidenced "great anxiety about speaking." The patients of Lebrun and Leleux (1985) and of Rousey et al. (1986) worried a lot about their loss of fluency. And Stewart and Grantham's patient (1993) found it extremely difficult to accept her new stuttering self.

9. Some people maintain that the onset of developmental stuttering is insiduous, whilst in acquired stuttering it is abrupt. This view also is debatable, as some children start to stutter rather suddenly, and adult-onset stuttering may come about gradually, as in the cases reported by Fleet and Heilman (1985), Godwin (1983, 2nd case) and Lebrun et al. (1983).

1O. Finally it is sometimes maintained that developmental and acquired stuttering do not respond to the same type(s) of therapy. However, an investigation conducted by Market et al. (1990) shows that generally the same therapeutical strategies are used for acquired as for developmental stuttering and with equal results.

It follows from all this that none of the descriptive features that have been said to demarcate developmental from acquired stuttering is fully reliable. Indeed, at times the symptomatology of adult-onset stuttering is undistinguishable from that of developmental stuttering.

Since the symptoms of acquired stuttering may so closely resemble those of developmental stuttering and since the two conditions appear to respond to the same types of treatment, one is justified in using the term *stuttering* to refer to either speech impairment. Indeed, as Ham (1995) humorously pointed out, rejecting the label *stuttering* when the acquired condition is meant, would amount to saying: "If it looks like a duck, walks like a

TABLE 1
Features which, despite contentions to the contrary, may be observed both in developmental and in acquired stuttering

1. Final sound repetitions	5. Monosymptomatology
2. Phrase repetitions	6. Adaptation effect
3. Function word repetitions	7. Secondary features
4. Imitative speech and oral reading as difficult as conversation	8. Concernedness
	9. Gradual or sudden onset
	10. Response to specific therapies

duck, quacks like a duck — let us call it a hummingbird with duck-like characteristics."

Forensic implications

The absence of reliable yardsticks by which to distinguish *symptomatologically* between developmental and acquired stuttering has forensic implications. As a matter of fact, after suffering cerebral trauma, an individual may evidence a stutter and maintain that he never stuttered before the accident. He may contend that the speech impediment results directly from the trauma. If someone else is responsible for the cerebral injury, the victim may claim financial compensation for his loss of fluency.

If one cannot demonstrate, on the basis of heteroanamnestic evidence, that the patient stuttered before the accident, his claim may have to be granted, since one cannot safely deduce from the mere symptomatology whether his stutter is developmental or acquired.

Objections to the labels "neurogenic stuttering" and "neurological stuttering"

While tolerating the use of *stuttering* to denote adult-onset dysfluency, a number of clinicians are in doubt as to the appropriateness of the adjectives *neurogenic* and *neurological* in this context. Some of these clinicians find that even if the stutter comes about shortly after cerebral damage, it is doubtful whether the loss of fluency is the direct consequence of the brain lesion. They reason that maybe the speech impediment results from the psychological stress brought on by the neurological deficits which the injury has caused.

Some take an even stronger view of the problem. In their opinion, stuttering that follows brain damage is always psychogenic. It occurs as a reaction to the distress which the neurological affection entails.

According to these clinicians, when both aphasia and stuttering are observed following brain injury, only the former is neurological in nature and the direct consequence of the organic damage. The latter is secondary. It results from the despondency caused by aphasia.

One of the exponents of this strong view was Goldstein (1948, p. 81) who contended that acquired stuttering in aphasics was "to be considered as an expression of the emotional distress into which the patient comes when he is aware of his (aphasic) defect." According to Goldstein, there is no essential difference between adult-onset stuttering and developmental stuttering: both are psychogenic in nature.

Golstein's opinion was echoed by Freund (1966, p. 145) who stated that "stuttering in the organically brain damaged represents a simple psychogenic reaction, whereby the psychological factor of awareness of a conflict between speech drive and speech accomplishment is of crucial importance."

However, the view that stuttering occurring in an adult who has suffered brain damage is always "a psychogenic overlay or reaction to the pathology" (Freund, 1966, p. 144), does not accord well with the observation that the loss of fluency may be the only significant deficit that is observed following cerebral injury. If the lesion, as in the case reported by Lebrun et al. (1990), has caused no aphasia, no intellectual loss and no motor or sensory handicap, why should the patient develop reactive stuttering?

There are also cases (see e.g. Lebrun et al., 1987b) in which following brain damage stuttering develops before aphasia. Obviously, in such cases, the stutter cannot be regarded as a psychological consequence of the language impairment.

The view that stuttering occurring in brain damaged people is always secondary to the neurological deficits, is also difficult to reconcile with those cases in which the patient starts stuttering as soon as he awakes from the initial coma. Under such circumstances, it can hardly be assumed that the stutter is a consequence of the distress experienced by the patient. If this were the case, the speech impediment would take some time to develop. It would come about after the patient has realized the severity of his predicament. It would not be present from the very beginning of wakefulness.

Again, when stuttering and aphasia are concomitant they may evolve differently. Aphasia may clear up while stuttering remains. Such dissociation also contradicts the view that stuttering is the consequence of aphasia.

Why should the stutter remain when the deficit that is supposed to have caused it, has disappeared?

Because they realized that there were cases in which stuttering co-occurring with aphasia could not possibly be considered to be reactive, Luchsinger and Arnold (1970, p. 701) proposed to distinguish between what they called *aphatisches Stottern* (aphasic stuttering) and *Stottern bei Aphasie* (stuttering in cases of aphasia). According to them, only the latter type of stuttering is psychogenic in nature and consequent upon the stress caused by aphasia. The former type is neurogenic in nature: like the aphasia it results directly from the cerebral lesion.

Unfortunately, Luchsinger and Arnold did not indicate how one can in actual clinical cases distinguish between the two types of adult-onset stuttering. Furthermore, like Goldstein, they failed to demonstrate that the affliction caused by aphasia could really result in stuttering.

Accordingly, the notion of reactive stuttering in cases with aphasia or other neurological deficits has never met with general acceptance. On the contrary, most clinicians consider that when stuttering comes about immediately or shortly after a patient has suffered demonstrable brain damage, the loss of fluency may be legitimately regarded as the direct consequence of the cerebral lesion and therefore be called *neurogenic* or *neurological stuttering*.

Neurogenic versus psychogenic acquired stuttering

There are patients in which stuttering appears not shortly after brain damage, but several months, or even a few years, later. In such cases, it may be questioned whether the stutter is caused by the cerebral injury. For instance, a patient described by Horner and Massey (1983) began to stutter some two years after a stroke which severely lesioned the right hemisphere both cortically and subcortically. In addition to involuntary sound, part-word, word and phrase repetitions, the patient evidenced reduced range of articulatory movements with mild speech sound imprecision, harsh voice quality and restricted pitch variation. Writing was somewhat micrographic. These various features had not been observed during language and speech skills assessment at one year post-stroke.

Horner and Massey proposed that in their patient a primarily subcortical impairment of speaking and writing developed gradually during the second year post-stroke as a late consequence of cortico-subcortical infarction with secondary enlargment of the right lateral ventricle.

However, because of the interval between the stroke and the appearance of stuttering, it might perhaps be argued that in Horner and Massey's case stuttering was reactive in nature. Admittedly, the patient had no aphasia but, due to left hemiplegia, was confined to a wheel chair. Somewhat contradictorily, his wife reported that he had become depressed and unconcerned. Therefore, the possibility of his stutter being a psychological consequence of his neurological condition cannot be totally excluded.

There are also patients with adult-onset stuttering who present with neurological deficits as well as with personality disorders. In such cases, it may be difficult to decide whether their dysfluent speech is on a par with their organic impairments or is on the contrary related to their psychological problems. A case in point has been reported by Deal and Doro (1987). Following traumatic head injury in his early twenties, their patient started to have grand mal seizures. During the subsequent 32 years he was in and out of hospitals, locked facilities, and jails. His medical records contained divergent diagnoses ranging from organic brain damage to schizophrenia to hysteria.

In his early fifties, this man began to experience spells during which he would lose consciousness for 2 to 30 minutes. These episodes were sometimes followed by transient mono- or hemiplegia, generally on the right.

One day he had a seizure with loss of consciousness. When he awoke, he had right hemiparesis and mutism. He could understand what he was said and could express himself in writing, but he appeared unable to speak. His speech suddenly returned about 1 month later.

During the ensuing 2 years, the patient was seen in different clinics for a fluctuating speech impairment. In addition, he complained of chest pains, headaches, sore throats, bouts of hemi- or monoparesis, and difficulty sleeping.

When finally seen by Deal and Doro, the patient had an intermittent severe stutter. He iterated initial syllables and, at times, also whole words. Some of the iterations lasted for over 30 seconds. Delivery was tachylalic. Verbal output was at times unintelligible.

During speech the patient maintained eye contact. His repetitions were not accompanied by grimaces or other involuntary movements. There was no change in speech quality when he read aloud, and there was no adaptation during repeated oral reading. When asked to mouth words without voice, he exhibited "what appeared to be the same pattern of repetition as when talking." On the other hand, he did not seem to be concerned about his dysfluencies. Alternating with his intermittent stutter he had transient motor difficulties which prevented him from walking.

Deal and Doro felt that their patient's stutter accorded rather well with the description of psychogenic stuttering of adult onset provided by Freund (1966, p. 139) under the name of *hysterical stuttering*. According to Freund, hysterical stuttering is stereotyped and omnipresent; the patient stutters to the same extent and produces the same types of dysfluencies in all oroverbal activities. There is no adaptation, i.e. no decrease in dysfluency on successive readings of the same text. Moreover, the patient maintains eye contact during speech and does not seem to worry about his stutter. Hysterical stuttering may be preceded by mutism.

Although the symptomatology of Deal and Doro's patient fits Freund's description of psychogenic adult-onset stuttering, one may wonder whether the stutter was not related to the brain injury which the patient had suffered in his twenties, and to the ensuing epilepsy.

It might perhaps be replied that if the speech impediment had been organically based, it would not have been intermittent. It should not be forgotten, however, that truly neurogenic speech disorders may very well come and go. There are Parkinsonian patients who suffer from paroxysmal mutism: there are times when they can speak and times when they are unable to utter words (Babinski et al., 1921) Other Parkinsonian patients have intermittent cluttering (Canter and Van Lancker, 1985). Paroxysmal dysarthria is observed in a number of people suffering from multiple sclerosis (Andermann et al., 1959; Espir et al., 1966; Miley and Forster, 1974). Ictal speech disorders are, by definition, intermittent (Lebrun, 1994). And in a case of progressive encephalopathy associated with chronic hemodialysis, acquired stuttering was seen to alternate with periods of fluent speech (Madison et al., 1977). In neurological patients, then, speech disorders including stuttering are not always permanent. Accordingly, intermittency is no guarantee that a stutter is psychogenic.

Neither is the occurrence of mutism before stuttering. If psychogenic stuttering can be preceded by an episode of mutism, as Freund asserts (1966, p. 140), so can neurogenic stuttering, as a case reported by Bijleveld et al. (1994) shows. Following what later proved to have been an ischemic accident in the left hemisphere, their patient suddenly lost the ability to express herself orally. A few days later, speech gradually returned but was dysfluent. The patient had never stuttered before. In addition to her dysfluency she showed soft pyramidal signs. Small infarcts could be neuroradiologically demonstrated in the left hemisphere. Her stutter was ascribed to the vascular brain damage she had suffered. Neurogenic stuttering, then, can be ushured in by speechlessness. Helm et al. (1980) reached the same conclusion on the basis of a case of stuttering following closed head injury.

It follows that in Deal and Doro's case the possibility of the stutter having been neurogenic cannot be discarded on the grounds that the dysfluency was intermittent and was initially preceded by an episode of mutism.

It should furthermore be remembered that at the time of examination Deal and Doro's patient had epilepsy of long standing. Presumably the patient had been given anticonvulsants repeatedly. Indeed, he was perhaps on permanent anticonvulsant medication. This may have played a role in the occurrence of dysfluent speech. As a matter of fact, several cases have been reported of stuttering appearing after chronic administration of anti-epileptic drugs (see page 126).

In view of all this, it must remain undecided whether Deal and Doro's case was an instance of psychogenic rather than of neurogenic stuttering. The more so as the case in various respects resembles the case reported by Dvorkin et al. (1987, pp. 214-215). Their patient had suffered from headache, vomiting and epilepsy since the age of 19. In addition he had had episodes of confusion, aggressiveness, impaired consciousness, mutism and blindness. At age 27 he was admitted to the hospital. He had cortical blindness and abnormal speech with paraphasias and a tendency to repeat phrases. While in hospital, he developed headache and became uncommunicative. A CT scan revealed extensive hypodensity in the posterior halves of both cerebral hemispheres. Despite the intermittency of the symptoms the organicity of the case was not questioned. It could very well be, therefore, that in Deal and Doro's case too acquired stuttering had been truly organic.

At times, acquired stuttering improves noticeably, or clears up completely, after a limited number of symptomatic speech therapy sessions. Rosenbek et al. (1978) described a right-handed patient who began to stutter four days after having suffered a stroke in the right parietal lobe. Two days later, a remedial programme of syllable-timed speech was started. After six sessions, the patient became fluent again.

Tippett and Siebens (1991) hold this case to be an instance of psychogenic acquired stuttering, on account of the quick responsiveness to treatment. They reason that if stuttering had been organically based, it would have taken much longer to eliminate it. Sapir and Aronson (1990) also regard rapid improvement as "the strongest evidence for psychogenic etiology."

However, it should be remembered that in neurological patients, speech disorders can be transitory. Short-lived dysarthria or dysfluency can occur as a consequence of transient ischemic disturbances or as a result of diaschisis or perilesional edema. If, as in the case reported by Rosenbek et al. (1978), therapy is started shortly after onset of the stutter and the speech impediment clears up after a few sessions, one can never be sure that the dysfluency was not in fact transitory. It may have subsided spontaneously, the therapy having had little or no influence on the process of recovery.

Accordingly, rapid improvement of the stutter cannot safely be regarded as strong evidence of its psychological origin.

Psychogenic stuttering

There are cases of adult-onset stuttering with no indication of cerebral injury but with obvious behavioural disorders and/or personality problems. In such cases, it seems reasonable to regard the speech impediment as psychogenic, particularly if the dysfluency can be given symbolic significance within the patient's emotional make-up. A case in point has been reported by Mahr and Leith (1992). Their patient was admitted to a psychiatric clinic because of an acute manic episode with psychosis. Past history was significant for conversion reactions involving left-sided paralysis and pelvic pain. After resolution of the manic behaviour, the patient was given psychotherapy. As traumatic memories of sexual abuse began to emerge, the patient started stuttering. The dysfluency was initially mild but became progressively more severe, so that it interfered with verbal intercourse. Outside the therapy room, however, the patient remained fluent. She was instructed to slow down her rate of speech and to enunciate each word clearly. This technique could be quickly mastered and it restored fluency.

In this case, there was no evidence of brain damage. Moreover, the stutter was observed only during therapy which involved calling up repressed memories of sexual abuse. There can be little doubt that the dysfluency was psychogenic in nature.

There are also cases with no evidence of brain damage and in which the stutter occurs in conjunction with a relatively sudden stressful situation that deeply affects the patient (Roth et al., 1989) or which the patient feels he cannot cope with. An instance of the latter was reported by Weiner (1981). His patient was a business man who started to stutter when his partner left the company they were running together. He had to handle the enterprise alone and did not feel up to the task.

Interestingly enough, the patient's symptomatology differed in several respects from Freund's characterization of hysterical stuttering. It began insiduously while, according to Freund, psychogenic adult-onset stuttering tends to begin suddenly. It was rather severe during business phone calls, less severe in conversation and inexistent in oral reading, whereas in Freund's opinion psychogenic dysfluency is not influenced by the nature of the verbal task. The patient was concerned about his stutter and sought medical advice, whilst in Freund's view the hysterical stutterer is not worried by his speech impediment.

Psychogenic adult-onset stuttering, then, has no uniform symptomatology. Its features may vary from patient to patient. Moreover, it is at times accompanied, or preceded, by other reactive behaviours. In 1961 Wallen reported the case of a service man who started to stutter rather severely during an acute anxiety reaction following a period of overwork under rather stressful conditions. In addition to stuttering, he evidenced frequent trembling and crying. These symptoms gradually tapered off to normal over a three weeks' period.

In war time, the advent of stuttering is sometimes observed in front-line soldiers who have been exposed to severe combat conditions (Peacher and Harris, 1946). This battle stress stuttering may be ushured in by mutism.

There are cases in which psychogenic stuttering is tentatively diagnosed because no cerebral damage or disease can be demonstrated. There are no obvious psychopathological disturbances and no known precipitant, but neurological and neuroradiological examinations fail to uncover any organic affection. Psychogeny is assumed by default. Leahy and Stewart (this volume) speak of *idiopathic stuttering* in such cases.

In psychogenic acquired stuttering, there may be repetitions as well as prolongations and blocks. Repetitions may involve sounds, part-words, words, or phrases. Sound and part-word repetitions occur primarily in word initial position, but may at times occur in word final position. Function words may be stuttered upon as frequently as content words.

Severity ranges from mild to very severe, with stuttering occurring in some cases on every (other) word. In addition, there may be involuntary muscle contractions resulting in such movements as grimacing, blinking, raising of the eyebrows, head jerking or shuffling of the feet.

Stuttering may affect all forms of oroverbal output or, on the contrary, be manifest in conversational speech but absent in oral reading, recitation, serial speech, and/or singing. Adaptation may or may not obtain.

Some patients express deep concern about their dysfluent speech while others appear rather unaffected.

In short, the clinical picture of psychogenic acquired stuttering is as varied as the clinical picture of neurological stuttering. Moreover, there is not a single verbal feature that infallibly distinguishes one condition from the other.

REPETITIONS

Repetition is one of the main symptoms of stuttering, both developmental and acquired. Indeed, in some cases, it is the only noticeable feature.

Repetitions that are characteristic of stuttering are involuntary. The patient repeats part of his utterance unintentionally. Even if he carefully monitors his speech, he may be unable to prevent repetitions from occurring.

Iterations and palilalias

Stutterers repeat sounds, part-words, words, phrases and/or sentences. Repetitions of short segments such as sounds, part-words and words are called *iterations*, whereas repetitions of phrases and sentences usually go by the name of *palilalias*. (Souques, 1908; Trénel and Crinon, 1912). When (polysyllabic) words are repeated many times in a row, some clinicians classify the repetitions as palilalias rather than as iterations.

Iterations of sounds and of part-words are usually more frequent in word initial than in word medial or word final position. Sound iterations tend to affect consonants, particularly occlusive consonants, more than vowels.

When a sound or part-word is repeated several times at one stretch, it is not always possible to ascertain by ear exactly how many times it is being produced. There is virtually no pause between the successive occurrences and their exact number is difficult to establish simply by listening to the speaker. Such iterations are often referred to as *clonic iterations*, *verbal cloni* or *logocloni*.

In palilalias, i.e. repetitions of phrases or sentences (Dupré and Le Savoureux, 1914), voice intensity may decrease progressively. Indeed, the end of the palilalia may be whispered. Palilalias with a whispered end are called *aphonic* (Marie et Levy, 1925).

The range of articulatory movements may also decrease gradually as the palilalia takes its course, so that the end is barely mumbled (Sterling, 1924). Palilalias, then, are not always produced as series of phonetically identical copies.

Some clinicians are of the opinion that in palilalias speech rate increases graduallly: the phrase or sentence is repeated with growing rapidity; a sort of festination takes place in the course of repetition (Boller et al., 1973, 1975; Helm, 1979; LaPointe and Horner, 1981). This opinion is based primarily on auditory perceptual impressions, and these may be quite misleading. The gradual decrease in voice volume and in range of articulatory movements together with a progressive shortening of the pauses between the successive productions of the repeated item may easily induce a sensation of festination. Actually, acoustic measurements are necessary to demonstrate gradual acceleration. So far only a limited number of measurements of word duration in palilalias have been published (e.g. by Kent and LaPointe, 1982; Ackermann et al., 1989) and they do not confirm the view that festination is a (constant) feature of palilalia.

Some patients with adult-onset stuttering produce only iterations, others only palilalias. Still others produce both iterations and palilalias. Iterations at times occur within palilalias.

Some clinicians distinguish sharply between verbal output that is replete with iterations but contains virtually no palilalias, and verbal output that contains palilalias but few, if any, iterations. According to them, only the former type of speech should be called *stuttering*; palilalic speech falls outside the scope of stuttering (Boller et al., 1975; Fleet and Heilman, 1985; Rosenbek, 1985).

However, the radical segregation of iterative stuttering from palilalic speech leaves one with a problem: how is verbal production to be classified that contains iterations as well as palilalias? A number of patients with extrapyramidal diseases repeat sounds, part-words, words, phrases and sometimes even sentences. Should their repetitive speech behaviour be regarded as stuttering or not?

Actually, the existence of patients who produce iterations as well as palilalias (Helm et al., 1978; Horner and Massey, 1983; LaPointe and Horner, 1981; Lebrun et al., 1986; Lebrun et al., 1991). suggests that the two types of repetitions are akin. They both seem to result from a lack of inhibition in motor speech sequencing abilities. In either case, a verbal segment that has already been externalized is unintentionally repeated. Therefore, it seems reasonable to regard both iterative and palilalic speech as forms of acquired stuttering.

In some patients, the successive repetitions of a given verbal segment never span over more than a respiratory cycle (Ackermann et al., 1989). In other patients, long repetitions are interspersed with respiratory pauses (Lebrun et al., 1986).

Deliberate versus unintentional repetitions

By definition, stuttering repetitions are unintentional. They occur in spite of the speaker. However, when analyzing speech samples, it is not always easy to ascertain whether a given repetition is voluntary or not. If a stutterer blocks on a word he may repeat the word(s) immediately preceding the one he has been unable to say, and attempt the problem word a second time.

In such a case, it is often uncertain whether word repetition takes place deliberately or not. Does the stutterer intentionally repeat the word(s) immediately preceding the one he has blocked on in the hope that this may facilitate the production of the problem word, or is the repetition as involuntary as the block itself?

Patients with aphasia or dysarthria at times seek to repair their speech errors. Their attempts at self-correction often entail part-word or word repetitions. Such repetitions should not be mistaken for stuttering for they are deliberate, whereas iterations that are characteristic of stuttering are unintentional.

In 1975 Farmer reported an investigation of stuttering repetitions in aphasic and nonaphasic brain damaged adults. As she did not specify how stuttering repetitions were distinguished from deliberate repetitions, the significance of her findings remains uncertain.

Iterations and palilalias versus perseverations

It is typical of iterations and palilalias that the speech elements which are repeated do not unduly take the place of other, more appropriate, elements. This trait to some extent distinguishes iterations and palilalias from perseverations. Perseverated speech elements are often substituted for the desired ones: speech segments which have been used earlier recur instead of the novel segments which should be used. Perseverations, then, often result in paraphasias. Iterations and palilalias never do.

Moreover, perseverations are not always immediate. Not infrequently, the undue recurrence takes place after a series of novel items has been produced. In iterations and palilalias, on the contrary, repetition tends to be immediate: it generally concerns the speech elements that have been produced last. In the rare cases where a novel speech fragment is inserted between the original item and its subsequent repetition(s), this fragment is usually short.

There is, however, one form of perseveration that closely resembles palilalia and final-position iteration or protraction, namely *continuation*. Continuous perseveration is the involuntary prolongation of an action beyond the point of completion. The patient fails to terminate his activity although the intended goal has been achieved. When a patient involuntarily repeats his utterance or the last part of his utterance, his behaviour can be viewed as a continuous perseveration: the speaking action is unintentionally prolonged beyond its normal end point. In this respect, it may be mentioned that some palilalic patients are also palipraxic (e.g. Foix and

Schulmann's patient, 1915): they uselessly repeat some of their gestures several times in succession.

NEUROGENIC STUTTERING

Neuropathology

Neurogenic stuttering may obtain after damage to the left or to the right hemisphere. The causal lesion may be cortical, subcortical or both. It is sometimes focal and sometimes diffuse. It may be vascular, traumatic, neoplastic, toxic or degenerative. Stuttering may also occur after brain concussion. Focal lesions are not invariably confined to a particular cerebral area, but may be found in various parts of the encephalon. Thus, the nature and the localization of the cerebral injury that causes acquired stuttering lack specificity.

Symptomatology

Neurogenic stuttering may begin suddenly or develop progressively. It may include repetitions, blocks and prolongations. In some cases repetitions are particularly conspicuous; in others, blocks and possibly prolongations predominate. As a rule, sound and part-word repetitions occur more frequently in word initial than in word medial or word final position.

Neurogenic stuttering may be present in all forms of oroverbal output including whispered speech and singing or, on the contrary, occur in some modalities of oral expression and not in others. For instance, it may affect conversational speech while (relatively) sparing serial speech, recitation, oral reading, and/or singing. At times, it is observed in only one form of speech. In Ackermann et al.'s case (1989), it occurred exclusively in repetition tasks: the patient could not help reproducing the stimulus words several times in a row instead of just once, as requested. He did not stutter otherwise.

Within affected oroverbal modalities, severity may be relatively constant or it may wax and wane. The patient may even experience occasional periods of fluent speech (Madison et al., 1977). Indeed, in some cases stuttering is observed only intermittently, e.g. at times of maximum plasma concentration level of anti-parkinsonian medication (Ackermann et al., 1989). Acquired stuttering, then, may be *paroxysmal*. On the other hand, when present, it may be so severe that the patient seems to be "gagging on his words" (Madison et al., 1977).

Involuntary movements may accompany the dysfluencies. These dyskinesias tend to affect the head more than other body parts. Eye contact may be maintained or be lost during stuttering.

In some patients, an adaptation effect is found. In others, repeated sequential reading of the same material does not reduce stuttering.

Neurogenic stuttering may be the initial symptom or the forerunner of a neurological disease, or it may (for some time) be the patient's sole complaint. In other cases, it occurs together with other neurological deficits.

Neurogenic stuttering following right brain damage

Following injury to the right hemisphere, an individual may present with acquired stuttering. The speech impediment may include repetitions, blocks and prolongations. Repetitions may affect sounds, part-words, words, phrases and even sentences. In addition, there may be some degree of cluttering: in the patient's hurried speech, a number of syllables are deleted or blended (Botez and Wertheim, 1959).

Usually there is no concomitant aphasia, but there may be some degree of dysarthria as well as aprosody. The patient may also have become avocalic, i.e. unable to sing. Written language may be impaired by afferent dysgraphia or neglect dyslexia-dysgraphia.

It would seem that neurogenic stuttering due to right brain damage is as frequent in right-handed as in left-handed or ambidextrous individuals.

In 1947 Eisenson, describing his experience with World War II casualties, noted that following right-sided cerebral injuries a number of right-handed soldiers were speechless for a short period of time. Upon regaining speech, a few of them evidenced stuttering.

Ever since a sizeable number of (right-handed) patients have been reported who showed acquired stuttering after right brain damage (Ardila and Lopez, 1986; Fleet and Heilman, 1985; Helm et al., 1978; Horner and Massey, 1983; Lebrun and Leleux, 1985; Lebrun et al., 1987b; Ludlow et al., 1987; Rétif et al., 1987; Rosenbek et al., 1978; Rosenfield et al., 1980; Soroker et al., 1990).

Neurogenic stuttering following left brain damage

Neurogenic stuttering may also result from injury to the left hemisphere. The causal lesion may lie in the frontal, parietal or temporal lobe. The stutter consists primarily of repetitions. There may in addition be blocks and prolongations, and the dysfluencies may be accompanied by (facial) dyskinesias.

Neurogenic stuttering consequent upon damage to the left hemisphere is often concomitant with aphasia. The latter may be of the motor or sensory type. When co-occurring with motor aphasia, stuttering may be an integral part of the aphasic syndrome or, on the contrary, be relatively independent of it. In the latter case, it may at some point clear up while aphasia remains, or it may persist even though aphasia subsides.

If stuttering co-occurs with transcortical motor aphasia (Ackermann et al., 1996; Jonas, 1981; Luria, 1970, p. 210; Rubens, 1975), it is usually inherent to the aphasic syndrome. The basic problem in transcortical motor aphasia is the unassisted self-generation of novel speech. Due to pathological inertia of the central mechanisms responsible for speech production, the patient tends to repeat what he has just said instead of uttering new words. The following translated excerpt from an audio-recording of a Dutch-speaking patient with transcortical motor aphasia may serve as an example: "I was admitted to-to the hospital and I was-was examined by a professor ... For treatment I had-I had to do-do exercices (What kind of exercises?) I should-I should be able to tell you but-but I c-can't".

In addition to repetitions of sounds, part-words and words, the disturbance in the dynamics of novel speech production may cause blocks or prolongations.

Initiating spontaneous or responsive speech may also be difficult in transcortical motor aphasia, because it implies a change from rest to self-generated oroverbal activity. The beginning of sentences may therefore be marked by efforts to get started. These efforts may result in the involuntary production of a speech sound. In some cases, this speech sound becomes stereotyped and is produced whenever the patient starts speaking. In addition, it may be uttered several times in succession, possibly in a clonic way. An epenthetic and at times stereotyped sound may also appear within utterances and be iterated. Generally, the dysfluencies of transcortical motor aphasia are not accompanied by dyskinesias.

Because it involves external support, imitative speech is easy in transcortical motor aphasia and free from iterations and blocks. For the same reason, reading aloud, if at all possible, is usually fluent.

Very occasionally transcortical motor aphasia with stuttering is observed after damage to the right hemisphere.

When it co-occurs with other forms of aphasia, stuttering does not appear to be an integral part of the aphasic syndrome. It is just a concomitant which evolves independently of the language disorder. Initially the patient may be totally speechless. When the ability to utter words is recovered, dysfluencies and aphasia may be observed. This type of evolution was briefly described by Gall early in the 19th century. In the fifth volume of his treatise *Sur les fonctions du cerveau et sur celles de chacune de ses parties* (1822–1825), the father of phrenology mentioned a nobleman who following traumatic damage to his left hemisphere lost the ability to speak. Eventually, his speechlessness subsided but he remained aphasic and stuttered slightly (p. 28)

In the last hundred years, a number of cases of acquired stuttering concomitant with aphasia have been reported, a.o. by Arend et al. (1962), Caplan (1972), Mazzucchi et al. (1981), Pick (1899), Schiller (1947), and Shtremel (1963).

Not all repetitions made by aphasics can be considered stutter-like dysfluencies, though. At times, (motor) aphasics *deliberately* repeat a sound or a part-word in an effort to utter a given target word. Similar efforts may also be observed in patients with speech apraxia (Lebrun, 1990). Such voluntary repetitions are not to be mistaken for stuttering.

Neurogenic stuttering following bilateral brain damage

In patients with closed head injuries or with skull fractures, there is usually more than one lesion. Indeed, both hemispheres may be injured, and the brainstem may have suffered too.

Patients with such lesions may evidence acquired stuttering (Marshall and Neuburger, 1987) possibly associated with aphasia (Andrews et al., 1972). The speech impediment may furthermore be concomitant with post-traumatic epilepsy. Both may not develop until a few years after the accident (Helm et al., 1980; Helm-Estabrooks, 1986; Quinn and Andrews, 1977). Neurogenic stuttering may also be consequent upon bilateral infarcts (Helm and Butler, 1977; Helm et al., 1978; Nowack and Stone, 1987).

Neurogenic stuttering following subcortical damage

Neurogenic stuttering is observed in a number of patients with extrapyramidal affections. Some of these patients produce impressive repetitions (Lebrun et al., 1986). There may also be blocks and prolongations. In Parkinsonian syndromes, blocks may be due to the inability to initiate the speech action, or to articulatory freezing, probably as a consequence of rigidity and hypokinesia. Not infrequently the stutter is associated with dysarthria. The speech impediment may be pronounced even if the non-verbal motricity is but midly affected.

At times, stuttering consists primarily of palilalias (Helm, 1979). Palilalic stuttering may be present in conversational speech, while sparing serial speech, oral reading and/or singing. When reciting an overlearned series, palilalic speakers may produce no palilalia but may tend to repeat the series all over again, once it is completed (Boller et al., 1973, 1975).

Palilalias and iterations may also occur in the speech of patients with lesions in the subcortical white matter (Lebrun et al., 1991).

Stuttering may result not only from neuropathology but also be occasioned by electrical or mechanical stimulation of brainstem structures. Electrostimulation or electrode displacement during stereotactic surgery performed on conscious patients may induce involuntary sound, part-word and word repetitions in responsive or serial speech (Andy and Bhatnagar, 1991; Ojemann, 1976; Ojemann et al., 1968; Ojemann and Ward, 1971; Schaltenbrand, 1975). At times, the repetition continues as long as the stimulation is maintained.

Stereotactic surgery performed with a view to alleviate dystonia may bring on durable stuttering possibly accompanied by dysarthria (Helm-Estabrooks, 1986).

Cortical versus subcortical stuttering

A number of authors have attempted to identify features distinguishing neurogenic stuttering due to a cortical lesion from neurogenic stuttering caused by subcortical dysfunction. Horner and Massey (1983) have proposed that syllable repetition is a "common symptom" in cortical while it is "the least common symptom" in subcortical stuttering. However, in the cases of subcortical stuttering reported by Cipolotti et al. (1988), Hertrich et al. (1993) and Lebrun et al. (1986) syllable repetition was the most frequent type of repetition.

Horner and Massey have also submitted that in cortical stuttering repetitions occur as often on content as on function words, whereas in subcortical stuttering they occur more often on content words. Yet, the extrapyramidal patient of Lebrun et al. (1986) stuttered more frequently on functors than on contentives. And Koller's fifth and sixth extrapyramidal patients (1983) stuttered as often on grammatical as on substantive words.

Horner and Massey have further claimed that in subcortical stuttering, repetitions occur much more frequently in self-formulated speech than in imitative speech, oral reading, recitation or singing. Koller has similarly maintained that extrapyramidal stuttering is rarely observed in imitative and choral speech. Yet, the extrapyramidal patient of Lebrun et al. (1986) stuttered as much in repetition tasks and in oral reading as in spontaneous speech. And the patient of Hertrich et al. (1993) produced more iterations in the repetition task than during conversation. As for Koller, he noted that singing did not abolish stuttering in two patients with extrapyramidal diseases.

Koller holds that a positive adaptation effect distinguishes extrapyramidal stuttering from stuttering secondary to stroke or trauma. But in Hertrich et al.'s extrapyramidal case (1993) successive readings of a text did not entail reduced stuttering.

In many respects, then, cortical and subcortical stuttering resemble each other, and they may be hard to tell apart on the sole basis of their linguistic features.

In ten veterans who exhibited chronic acquired stuttering following penetrating missile wounds Ludlow et al. (1987) found "predominantly unilateral lesions of the subcortical pyramidal and extrapyramidal systems". It could therefore be the case that neurogenic stuttering of subcortical origin tends to be more persistent than neurogenic stuttering due to pure cortical lesions.

Neurogenic stuttering following diffuse brain damage

Not only focal but also diffuse brain damage may bring on neurological stuttering. At times, only one system is diffusely affected, as in motor neurone disease (Lebrun et al., 1983). In other cases, the causal affection is polysystemic.

There are patients in whom stuttering and aphasia develop concurrently as initial symptoms of degenerative dementia (Craenhals et al., 1990; Kirshner et al., 1984). In other patients with degenerative dementia, stuttering appears at a later stage (Helm-Estabrooks et al., 1989; Lebrun et al., 1987b; Rétif et al., 1988). At advanced stages, verbal output may consist for the most part of reiterated syllables (Lebrun et al., 1987a).

Neurogenic stuttering may be consequent upon intoxication of the central nervous system. Patients who have undergone prolonged periods of hemodialysis for kidney dysfunction may show various neurological symptoms including acquired stuttering and dysarthria (Rosenbek et al., 1975). The stutter may be initially severe and decrease as the patient's general condition worsens (Madison et al., 1977).

Several patients have been reported in whom adult-onset stuttering developed after (chronic) administration of anticonvulsants (Attanasio, 1987; Helm et al., 1980; Helm-Estabrooks, 1986; Nowak and Stone, 1987). Changing anti-epileptic medication may result in an improvement of the stutter (McClean and McLean, 1985). Stuttering, then, may be a sign of the neurotoxicity of some anticonvulsive drugs.

Acquired stuttering has also been observed after administration of neuroleptics (Friedman, 1990; Guthrie and Grunhaus, 1990; McCall, 1994; Nurnberg and Greenwald, 1981; Quader, 1977; Thomas et al., 1994). Withdrawal of the drug or dose reduction usually results in the subsidence of the dysfluency. Stuttering may also occur following tranquilizer abuse (Rentschler et al., 1984).

Interestingly enough, several of the published cases of acquired stuttering that have been given a psychological explanation had a history of epilepsy, depression or drug abuse, e.g. Deal, 1982 : heroin addiction, chronic depression, attempted suicide (with barbiturates?); LaPointe and Horner, 1981: suspected phenobarbital addiction, barbiturate abuse; Tippet and Siebens, 1991: anticonvulsants, depression. It could be that in these cases, or in some of them, the narcotics, the anticonvulsants, the psychotropic drugs and/or the antidepressants were, at least in part, responsible for the emergence of adult-onset stuttering. In other words, these cases, rather than being psychogenic, may have been due to intoxication of the central nervous system.

Neurogenic stuttering in the context of migraine

Persistent acquired stuttering was observed by Stewart and Grantham, (1993) in a young adult with a history of migraine. One morning the patient "awoke with a migraine headache and a pain over her right eye (This differed from her usual symptoms, because in previous episodes the migraine had been focussed over the left eye.) Later that day she noted a bilateral tremor. Her headache disappeared the following day, but the tremor remained for about 10 days on the left and slightly longer in her right hand. At that time she reported having to concentrate on sentence formation, but speech production and fluency were normal. One week later she reported having difficulty with speech fluency at a severe level (i.e. non-fluent on every word)."

Because subsequent neurological and neuroradiological examinations failed to uncover any anomaly apart from the persisting stutter, Leahy and Stewart (this volume pp. 139-152) construe this case as an instance of psychogenic stuttering. However, since stuttering appeared following a migrainous episode with tremor and sentence formation difficulties, it may well have been organically based, as in the case of stuttering associated with vascular headache reported by Rosenfield (1972).

Prolongations in cerebellar dysarthria

In his description of neurogenic stuttering, Canter (1971) included the phonemic prolongations which occur in the speech of many patients with cerebellar dysarthria. Such prolongations are due to the dysmetria which often affects speech movements when the cerebellum is lesioned. As Canter himself pointed out, sound protractions result from the patient's inability to properly gauge the speed and force of voluntary movements. This inability gives rise not only to phonemic lengthenings, but also to phonemic shortenings as well as to sudden and unexpected changes in pitch and in voice intensity. Such dysmetric delivery is vastly different from stuttering. Accordingly, it does not seem desirable to include cerebellar phonemic protractions in a description of neurogenic stuttering.

Prognosis and therapy

In a number of cases neurogenic stuttering improves spontaneously. It may even clear up completely. In other cases, reduction or subsidence of the dysfluencies is observed following treatment for concomitant or intercurrent pathologies, such as surgery for the alleviation of carotid stenosis (Donnan, 1979), or the removal of a tumour (Shtremel, 1963), or as secondary epilepsy becomes pharmacologically controlled (Baratz and Mesulam, 1980). In a few patients, electric self-stimulation by means of chronic electrodes implanted in the thalamus for the alleviation of intractable pain resulted in an marked improvement of speech fluency (Andy and Bhatnagar, 1992). In some cases, anti-parkinsonian medication similarly brings about a reduction of stuttering (Koller, 1983).

At times, the speech impediment is tackled directly. Often the patient is taught to speak slowly, in a syllable-timed way (Market et al., 1990; Stewart and Rowley, 1996). Occasionally, mechanical support is needed to achieve rate control. Helm (1979) devised a pacing board which had six multicoloured squares separated by raised wooden dividers. The patient learned to tap his finger from square to square while speaking in a syllable-by-syllable manner.

In some cases, delivery control can be achieved using a delayed auditory feedback device (Downie et al., 1981; McClean and McLean, 1985; Marshall and Neuburger, 1987) or an auditory feedback masking device (Dewar et al., 1979). However, stuttering tends to return when auditory feedback ceases to be delayed or masked. Moreover, the beneficial influence of the device may wear off with time.

In a patient with acquired stuttering following repeated bilateral strokes,

Helm and Butler (1977) found that syllable-timed speech increased the dysfluencies (which were primarily blocks), while transcutaneous nerve stimulation applied to the bicipital groove of the left arm noticeably reduced them.

Some patients derive benefit from biofeedback and relaxation therapy, while others do not (Helm-Estabrooks, 1986).

Pathophysiology of neurogenic stuttering

At the end of the 19th century, Pick (1899) already pointed out that the lesions associated with neurogenic stuttering are not invariably confined to a particular cerebral area but, on the contrary, may be found in various parts of the brain. In some patients cerebral damage is primarily, if not exclusively, cortical, whereas in others it mainly affects subcortical structures. In this respect, neurological stuttering resembles hypertonia. Muscle tone can become exaggerated following cortical as well as following subcortical injuries. In the former case, the increased tonus is generally called *spasticity*, in the latter, *rigidity*. Spasticity is usually considered to result from the release of subcortical structures from cortical control, whereas rigidity appears to be due directly to subcortical, in particular extrapyramidal, damage. In either case, the disorder of muscle tone is ascribable to an unbridled or abnormal functioning of structures in or near the brainstem.

Could it be that neurogenic stuttering also results from inadequate intervention of subcortical areas? Are there indications that the extrapyramidal system might be involved in the production of involuntary repetitions, blocks and prolongations?

As was shown above, there is a ample evidence that extrapyramidal diseases can entail stuttering. Mechanical or electrical stimulation of diencephalic structures may also induce involuntary repetitions. Stereotactic surgery occasionally results in acquired stuttering. Administration of psychotropic drugs may induce a number of extrapyramidal side-effects including stuttering. Conversely, therapeutic electrostimulation of the thalami and anti-parkinsonian medication at times bring about a reduction of verbal dysfluency.

It appears, then, that pathological changes in the basal ganglia or in the thalami may cause stuttering. Since stuttering may also occur following cortical damage or damage to the fibres linking the cortex with the brainstem (Lebrun et al., 1991), it may be hypothesized that under normal circumstances, the extrapyramidal system takes part in speech production under the control of the cortex. As in other (over)learned motor behaviours, the basal ganglia probably take care of the routinized or less deliberate aspects of the action. In mother-tongue use, these presumably include articulation, rate of delivery and semi-conscious accompanying nonverbal body movements.

If the extrapyramidal system is damaged, the smooth production of ongoing speech is disrupted: there occur reverberations and stoppages. If following injury, the cortex is no longer able to control the extrapyramidal system, the latter tends to run wild, and similar speech disturbances are observed. There is some sense, then, in ascribing neurogenic stuttering to a dysfunction or hyperfunction of the extrapyramidal system.

The physiopathology of psychogenic stuttering

But what about psychogenic stuttering? How can adult-onset stuttering be explained when it is observed in the absence of demonstrable brain damage? Can the extrapyramidal system be again implicated? It would seem that it can. As a matter of fact, this system lies close to, and is intimately linked up with, the limbic system, which is the main neural substrate of emotional life. Affects and indeed psychic stress may, therefore, easily influence, and even disturb, the extrapyramidal system. Accordingly, it is not unthinkable that emotions and psychological problems may act upon the basal ganglia and the thalami and induce stuttering.

Developmental stuttering

Since the symptomatology of adult-onset stuttering often resembles that of developmental stuttering, the question arises whether the two forms of stuttering might not have a common etiological basis. Could there be some link between the extrapyramidal system and developmental stuttering? Could it be the case that in developmental stutterers cortical control over extrapyramidal participation in speech production is inadequate?

It has often been noted that developmental stutterers speak more fluently, indeed some of them become completely fluent, when they deliberately depart from their usual speaking manner. If a stutterer imitates someone else's way of speaking or purposely changes his voice or accent (for instance, in order to play a role on the stage), or simply voluntarily speaks in a protracted or sing-song way, he very often ceases to stutter.

It would seem that deliberate departure from one's usual speech pattern results in greater fluency in stutterers precisely because such departure implies a greater cortical participation in the speech action. In all likelihood, intentionally altering one's usual speech pattern means de-automatizing speech production and thus reducing the intervention of the extrapyramidal system.

In this view, therapies which aim at having the patient use a controlled, well-paced, syllable-timed speech are in fact therapies which strive to enhance cortical and to diminish extrapyramidal participation in speech production.

It has repeatedly been observed that developmental stutterers tend not to stutter when they use a foreign language which they are not fully conversant with. This fact impressed Jean-Marc Itard so much that in his *Mémoire sur le Bégaiement* (1817) he recommended that young stutterers should be given a foreign nanny, so that they would be obliged to use a foreign language and thus be able to speak without stuttering.

Probably when he uses a foreign language which he does not fully master, a speaker has to rely mostly, if not exclusively, on his cortex. Few, if any, routines have been created and extrapyramidal intervention is minimal. Therefore, if the extrapyramidal system is selectively lesioned, the foreign language will be less impaired than the mother-tongue, as a case reported by Aglioti and Fabbro (1993) shows.

If a stutterer has only a limited command of a foreign language, he is likely not to stutter in that language, as the use of it is of necessity dependent almost exclusively on the cortex. However, if through practice and exercises he eventually achieves a good command of the foreign language, stuttering is likely to manifest itself also in that language, probably because a number of routines have now come about, thus increasing the part played by the extrapyramidal system in the speech act.

The dramatic improvement that is often observed when developmental stutterers increase cortical participation in speech production reminds one of the improvement in gait which can be achieved by a number of Parkinsonian patients when they de-automatize walking movements, e.g. by displacing their feet in such a way that they never touch the lines separating the flag-stones making up the pavement. By stepping in a controlled, "unnatural" way, they walk better, most probably because they reduce extrapyramidal participation in the action.

There are further similarities between developmental stuttering and extrapyramidal diseases. In fact, developmental stuttering usually increases under conditions of stress, but it may disappear temporarily under the influence of a strong affect. When they are in a rage, stutterers sometimes vociferate without stuttering. Similar observations may be made in Parkinsonian patients who have speech difficulties. When stressed they find it even more difficult to express themselves. However, if they are outraged, they may speak fluently for a while (Babinski et al., 1921)

In an number of cases, the severity of developmental stuttering is variable: at times, patients speak fluently or nearly so, at other times they stumble over their words. A comparable on-and-off phenomenon can be observed in some Parkinsonian patients whose ability to utter words fluctuates considerably.

The severity of developmental stuttering may also vary with the propositionality of the verbal output, rote speech being less disturbed than expository speech. Likewise, patients with extrapyramidal diseases may be able to recite aloud while they are unable to discourse (Babinski et al., 1921).

These various observations point to a link between developmental stuttering and the extrapyramidal system. Indeed, they suggest that developmental stuttering results from a lack of adequate dominance of the cortex over the brainstem.

Degenerative diseases which tend to affect lower parts of the central nervous system before the cortex, may conceivably put an end to developmental stuttering by reducing, or intefering with, the activity of the extrapyramidal system. It is no wonder, therefore, that Miller (1985) should have observed cessation of developmental stuttering in an adult with progressive multiple scerosis.

On occasion, developmental stutterers are encountered who, having suffered brain trauma with loss of consciousness, speak fluently when they awake from the coma (Cooper, 1983; Helm-Estabrooks et al., 1986). In such cases verbal fluency is probably the consequence of the blow sustained by the brainstem during the accident: having suffered concussion or contusion (as attested by the coma period), subcortical structures are less able to resist cortical influence and control.

Again, surgery for the resection of a cerebral tumour or the clipping of a cerebral aneurysm may be followed by a lasting remission of developmental stuttering (Jones, 1966). Cortectomies for the alleviation of intractable epilepsy may have the same effect (Guillaume et al., 1957; Mazars et al., 1970). Presumably the structural changes brought about surgically improve the functioning of the cortex, enabling the latter to better control subcortical speech mechanisms and thus to avoid stuttering.

CONCLUSION

All this seems to support the conclusion that acquired and developmental stuttering are not radically different. Both appear to result from excessive autonomy of the brainstem in regard to the cortex. In stutterers, subcortical structures involved in speech production are insufficiently subordinated to, or regulated by, the grey matter of the cerebral hemispheres. The hierarchical organisation of the brain mechanisms responsible for oroverbal (self-)expression is deficient. In developmental stutterers this organization fails to be established, in subjects with acquired stuttering it is disrupted.

In the past, some clinicians upheld the view that stuttering was due to a lack of adequate dominance of one cerebral hemisphere over the other. It looks as if the speech impediment actually has its origin in an insufficient dominance of the cortex over subcortical structures. In 1865 Paul Broca proclaimed that we speak from the left hemisphere. We may have to add: But we stutter from the brainstem.

REFERENCES

ACKERMANN H., HERTRICH I., ZIEGLER W., BITZER M., BIEN S. (1996) Acquired dysfluencies following infarction of the left mesiofrontal cortex. *Aphasiology* 10: 409-417.

ACKERMANN H., ZIEGLER W., OERTEL W. (1989) Palilalia as a symptom of levodopa induced hyperkinesia in Parkinson's disease. *Journal of Neurology, Neurosurgery, and Psychiatry* 52: 805-807.

AGLIOTI S., FABBRO F. (1993) Paradoxical selective recovery in a bilingual aphasic following subcortical lesions. *NeuroReport* 4: 1359-1362.

ANDERMANN F., COSGROVE J., LLOYD-SMITH D., WALTERS A. (1959) Paroxysmal dysarthria and ataxia in multiple sclerosis. *Neurology* 9: 211-215.

ANDREWS G., QUINN P., SORBY W. (1972) Stuttering: An investigation into cerebral dominance for speech. *Journal of Neurology, Neurosurgery and Psychiatry* 35: 414-418.

ANDY O., BHATNAGAR S. (1991) Thalamic-induced stuttering. *Journal of Speech and Hearing Research* 34: 796-800.

ANDY O., BHATNAGAR S. (1992) Stuttering acquired from subcortical pathologies and its alleviation from thalamic perturbation. *Brain and Language* 42: 385-401.

ARDILA A., LOPEZ M. (1986) Stuttering associated with right hemisphere lesion. *Brain and Language* 27: 239-246.

AREND R., HANDZEL L., WEISS B. (1962) Dysphatic stuttering. *Folia Phoniatrica* 14: 55-56.

ATTANASIO J. (1987) A case of late-onset or acquired stuttering in adult life. *Journal of Fluency Disorders* 12: 287-290.

BABINSKI J., JARKOWSKI B., PLICHET X. (1921) Kinésie paradoxale. Mutisme parkinsonien. *Revue Neurologique* 28:1266-1270.

BARATZ R., MESULAM M. (1981) Adult-onset stuttering treated with anticonvulsants. *Archives of Neurology* 38: 132.

BHATNAGAR S., ANDY O. (1989) Alleviation of acquired stuttering with human centre-median thalamic stimulation. *Journal of Neurology, Neurosurgery, and Psychiatry* 52: 1182-1184.

BIJLEVELD H., LEBRUN Y., VAN DONGEN H. (1994) A case of acquired stuttering. *Folia Phoniatrica et Logopedica* 46: 250-253.

BLOODSTEIN O., GANTWERK B. (1967) Grammatical function in relation to stuttering in young children. *Journal of Speech and Hearing Research* 10: 786-789.

BOLLER F., ALBERT M., DENES F. (1975) Palilalia. *British Journal of Disorders of Communication* 10: 92-97.

BOLLER F., BOLLER M., DENES G., TIMBERLAKE W., ZIEPER I., ALBERT M. (1973) Familial palilalia. *Neurology* 23: 1117-1125.

BOTEZ M., WERTHEIM N. (1959) Expressive aphasia and amusia. *Brain* 82: 186-202.

BROCA P. (1865) Du siège de la faculté du langage articulé. *Bulletin de la Société d'Anthropologie* 6: 377-393.

BROOKSHIRE R. (1989) A dramatic response to behavior modification by a patient with rapid onset of dysfluent speech. In HELM-ESTABROOKS N., ATEN J. (eds.) *Difficult diagnoses in adult communication disorders*. Boston, College Hill Press: 3-12.

CANTER G. (1971) Observations on neurogenic stuttering: A contribution to differential diagnosis. *British Journal of Disorders of Communication* 6: 139-143.

CANTER G., VAN LANCKER D. (1985) Disturbances of the temporal organization of speech following bilateral thalamic surgery in a patient with Parkinson's disease. *Journal of Communication Disorders* 18: 329-349.

CAPLAN L. (1972) An investigation of stuttering-like speech in adult dysphasic subjects. *The Journal of the South African Speech and Hearing Association* 19: 52-66.

CIPOLOTTI L., BISIACCHI P., DENES G., GALLO A. (1988) Acquired stuttering: A motor programming disorder? *European Neurology* 28: 321-325.

COOPER E. (1983) A brain-stem contusion and fluency: Vicki's story. *Journal of Fluency Disorders* 8: 269-274.

CRAENHALS A., RAISON-VAN RUYMBEKE A., RECTEM D., SERON X., LATERRE E. (1990) Is slowly progressive aphasia actually a new clinical entity? *Aphasiology* 4: 485-510.

CULATTA R., LEEPER L. (1988) Dysfluency isn't always stuttering. *Journal of Speech and Hearing Disorders* 53: 486-487.

DEAL J. (1982) Sudden onset of stuttering: A case report. *Journal of Speech and Hearing Disorders* 47: 301-303.

DEAL J., DORO J. (1987) Episodic hysterical stuttering. *Journal of Speech and Hearing Disorders* 52: 299-300.

DEWAR A., DEWAR A., AUSTIN W., BRASH H. (1979) The long term use of an automatically triggered auditory feedback masking device in the treatment of stuttering. *British journal of Disorders of Communication* 14: 219-229.

DONNAN G. (1979) Stuttering as a manifestation of stroke. *Medical Journal of Australia* 1: 44-45.

DOWNIE A., LOW J., LINDSAY D. (1981) Speech disorders in Parkinsonism; use of delayed auditory feedback in selected cases. *Journal of Neurology, Neurosurgery, and Psychiatry* 44: 852-853.

DUPRE E., LE SAVOUREUX X. (1914) Palilalie chez une pseudo-bulbaire. *Revue Neurologique* 25: 453-456.

DVORKIN G., ANDERMAN F., CARPENTER S., MELANSON D;, VERRET S., JACOB J., SHERWIN A., BEKHOR S., LUGARESI E., SACKELLARES C., WILLOUGHBY J., MACGREGOR D. (1987) Classical migraine, intractable epilepsy and multiple strokes: A syndrome related to mitochondrial encephalomyopathy. In ANDERMAN F., LUGARESI E. (eds.) *Migraine and epilepsy*, Boston, Butterworths: 203-232.

EISENSON J. (1947) Aphasics: Observations and tentative conclusions. *Journal of Speech Disorders* 12: 290-292.

ESPIR M., WATKINS S., SMITH H. (1966) Paroxysmal dysarthria and other transient neurological disturbances in disseminated sclerosis. *Journal of Neurology, Neurosurgery, and Psychiatry* 29: 323-330.

FARMER A. (1975) Stuttering repetitions in aphasic and nonaphasic brain damaged adults. *Cortex* 11: 391-396.

FLEET W., HEILMAN K. (1985) Acquired stuttering from a right hemisphere lesion in a right-hander. *Neurology* 35: 1343-1346.

FOIX C., SCHULMANN X. (1915) A propos d'un cas de palilalie. *Revue Neurologique* 26: 779-782.

FREUND H. (1966) *Psychopathology and the problems of stuttering*. Springfield, Thomas.

FRIEDMAN E. (1990) Fluoxetine and stuttering. *Journal of Clinical Psychiatry* 51: 310.

GALL F.(1822-25) *Sur les fonctions du cerveau et sur celles de chacune de ses parties*. Paris, Baillière.

GODWIN R. (1984) Sudden onset of stuttering in adults: Three case studies. In EDWARDS M. (ed.) XIX Congress of the International Association of Logopedics and Phoniatrics (Edinburgh, August 1983) London, College of Speech Therapists: II, 597-602.

GOLDSTEIN K. (1948) *Language and language disturbances*. New York, Grune & Stratton.

GUILLAUME J., MAZARS G., MAZARS Y. (1957) Intermédiaire épileptique dans certains types de bégaiement. *Revue Neurologique* 96: 59-61.

GUTHRIE S., GRUNHAUS L. (1990) Fluoxetine-induced stuttering. *Journal of Clinical Psychiatry* 51: 85.

HAM R. (1995) Review of "Treating disordered speech motor control: For clinicians by clinicians". *Journal of Fluency Disorders* 20: 77-80.

HELM N. (1979) Management of palilalia with a pacing board. *Journal of Speech and Hearing Disorders* 44: 350-353.

HELM N., BUTLER R. (1977) Transcutaneous nerve stimulation in acquired speech disorder. *The Lancet* for Dec. 3: 1177-1178.

HELM N., BUTLER R., BENSON D. (1978) Acquired stuttering. *Neurology* 28: 1159-1165.

HELM N., BUTLER R., CANTER G. (1980) Neurogenic acquired stuttering. *Journal of Fluency Disorders* 5: 269-279.

HELM-ESTABROOKS N. (1986) Diagnosis and management of neurogenic stuttering in adults. In ST. LOUIS K. (ed.) *The atypical stutterer*. Orlando, Academic Press: 193-217.

HELM-ESTABROOKS N., NICHOLAS M., MORGAN A. (1989) Fifty-two-year old male with stammering speech - CVA? In HELM-ESTABROOKS N., ATEN J. (eds.) *Difficult diagnoses in adult communication disorders*. Boston, College Hill Press: 205-216.

HELM-ESTABROOKS N., YEO R., GESCHWIND N., FREEDMAN M., WEINSTEIN C. (1986) Stuttering. Disappearance and reappearance with acquired brain lesions. *Neurology* 36: 1109-1112.

HERTRICH I., ACKERMANN H., ZIEGLER W., KASCHEL R. (1993) Speech iterations in Parkinsonism: A case study. *Aphasiology* 7: 395-406.

HORNER J., MASSEY W. (1983) Progressive dysfluency associated with right hemisphere disease. *Brain and Language* 18: 71-85.

ITARD J. (1817) Mémoire sur le bégaiement. *Journal Universel des Sciences Médicales* 7: 129-144.

JONAS S. (1981) The supplementary motor region and speech emission. *Journal of Communication Disorders* 14: 349-373.

JONES R. (1966) Observations on stammering after localized cerebral injury. *Journal of Neurology, Neurosurgery, and Psychiatry* 29: 192-195.

KENT R., LAPOINTE L. (1982) Acoustic properties of pathologic reiterative utterances: A case study of palilalia. *Journal of Speech and Hearing Research* 25: 95-99.

KIRSHNER H., WEBB W., KELLY M., WELLS C. (1984) Language disturbance. An initial symptom of cortical degeneration and dementia. *Archives of Neurology* 41: 491-496.

KOLLER W. (1983) Dysfluency (stuttering) in extrapyramidal disease. *Archives of Neurology* 40: 175-177.

LAPOINTE L., HORNER J. (1981) Palilalia: A descriptive study of pathological reiterative utterances. *Journal of Speech and Hearing Disorders* 46: 34-38.
LEBRUN Y. (1990) Apraxia of speech: A critical review. *Journal of Neurolinguistics* 5: 379-406.
LEBRUN Y. (1994) Ictal verbal behaviour: A review. *Seizure* 3: 45-54.
LEBRUN Y., BIJLEVELD H., ROUSSEAU J. (1990) A case of persistent neurogenic stuttering following a missile wound. *Journal of Fluency Disorders* 15: 251-258.
LEBRUN Y., DEVREUX F., ROUSSEAU J. (1986) Language and speech in a patient with a clinical diagnosis of progressive supranuclear palsy. *Brain and Language* 27: 247-256.
LEBRUN Y., DEVREUX F., ROUSSEAU J. (1987a) Disorders of communicative behavior in degenerative dementia. *Folia Phoniatrica* 39: 1-8.
LEBRUN Y., DEVREUX F., ROUSSEAU J. (1991) Language and speech in a case of perivenous encephalomyelitis. *Aphasiology* 5: 87-95.
LEBRUN Y., LELEUX C. (1985) Acquired stuttering following right brain damage in dextrals. *Journal of Fluency Disorders* 10: 137-141.
LEBRUN Y., LELEUX C., RETIF J. (1987b) Neurogenic stuttering. *Acta Neurochirurgica* 85: 103-109.
LEBRUN Y., RETIF J., KAISER G. (1983) Acquired stuttering as a forerunner of motor-neuron disease. *Journal of Fluency Disorders* 8: 161-167.
LEBRUN Y., VAN BORSEL J. (1990) Final sound repetitions. *Journal of Fluency Disorders* 15: 107-113.
LUCHSINGER R., ARNOLD G. (1970) *Handbuch der Stimm- und Sprachheilkunde*. Wien, Springer-Verlag.
LUDLOW C., ROSENBERG J., SALAZAR A., GRAFMAN J., SMUTOK M. (1987) Site of penetrating brain lesions causing chronic acquired stuttering. *Annals of Neurology* 22: 60-66.
LURIA A. (1970) *Traumatic aphasia*. The Hague, Mouton.
McCALL W. (1994) Sertraline-induced stuttering. *Journal of Clinical Psychiatry* 55: 316.
McCLEAN M., McLEAN A. (1985) Case report of stuttering acquired in association with phenytoin use for post head injury seizures. *Journal of Fluency Disorders* 10: 241-256.
MADISON D., BAEHER E., BAZELL M., HARTMANN K., MAHURKAR S., DUNEA G. (1977) Communicative and cognitive deterioration in dialysis dementia. Two case studies. *Journal of Speech and Hearing Disorders* 42: 238-246.
MAHR G., LEITH W. (1992) Psychogenic stuttering of adult onset. *Journal of Speech and Hearing Research* 35: 283-286.
MANDERS E., BASTIJNS P. (1988) Sudden recovery from stuttering after an epileptic attack: A case report. *Journal of Fluency Disorders* 13: 421-425.
MARIE P., LEVY G. (1925) Un singulier trouble de la parole: La palilalie. *Le Monde Médical* 35: 329-344.
MARKET K., MONTAGUE J., BUFFALO M., DRUMMOND S. (1990) Acquired stuttering. *Journal of Fluency Disorders* 15: 21-33.
MARSHALL R., NEUBURGER S. (1987) Effects of delayed auditory feedback on acquired stuttering following head injury. *Journal of Fluency Disorders* 12: 355-365.
MAZARS G., HECAEN H., TZAVARAS A., MERIENNE L. (1970) Contribution à la chirurgie de certains bégaiements et à la compréhension de leur physiopathologie. *Revue Neurologique* 122: 213-220.
MAZZUCHI A., MORETTI G., CARPEGGIANI P., PARMA M., PAINI P. (1981) Clinical observations on acquired stuttering. *British Journal of Disorders of Communication* 16: 19-30.
MILEY C., FORSTER R. (1974) Paroxysmal signs and symptoms in multiple sclerosis. *Neurology* 24: 458-461.

MILLER A. (1985) Cessation of stuttering with progressive multiple sclerosis. *Neurology* 35: 1341-1343.

MOWRER D. (1987) Repetition of final consonants in the speech of a young child. *Journal of Speech and Hearing Disorders* 52: 174-178.

NOWACK W., STONE E. (1987) Acquired stuttering and bilateral cerebral disease. *Journal of Fluency Disorders* 12: 141-146.

NURNBERG H., GREENWALD B. (1981) Stuttering: An unusual side effect of phenothiazines. *American Journal of Psychiatry* 138: 386-387.

OJEMANN G. (1976) Subcortical language mechanisms. In WHITAKER H. (ed.) *Studies in neurolinguistics*, vol. 1. New York, Academic Press: 103-138.

OJEMANN G., FEDIO P., VAN BUREN J. (1968) Anomia from pulvinar and subcortical parietal stimulation. *Brain* 91: 99-116.

OJEMANN G., WARD A. (1971) Speech representation in ventrolateral thalamus. *Brain* 94: 669-680.

PEACHER W., HARRIS W. (1946) Speech disorders in World War II: VII Stuttering. *Journal of Speech Disorders* 11: 303-308.

PICK A. (1899) Ueber das sogenannte aphatische Stottern als Symptom verschiedenörtlich localisirter cerebraler Herdaffectionen. *Archiv für Psychiatrie und Nervenkrankheiten* 32: 447-469.

QUADER S. (1977) Dysarthria: An unusual side effect of tricyclic antidepressants. *Brithsh Medical Journal* 9: 97.

QUINN P., ANDREWS G. (1977) Neurological stuttering - A clinical entity? *Journal of Neurology, Neurosurgery, and Psychiatry* 40: 699-701.

RENTSCHLER G., DRIVER L., CALLAWAY E. (1984) The onset of stuttering following drug overdose. *Journal of Fluency Disorders* 9: 265-284.

RETIF J., LEBRUN Y., RAES J., LELEUX C. (1987) Le bégaiement neurogène: Rôle d'une lésion hémisphérique droite dans sa genèse chez le droitier. In *Congrès de Psychiatrie et de Neurologie de Langue Française*, 85e session: 443-446.

RETIF J., LELEUX C., LEBRUN Y., DE MOL J. (1988) Forme de la maladie d'Alzheimer à début aphasique pur. In *Congrès de Psychiatrie et de Neurologie de Langue Française*. 86e Session: 275-287.

ROSENBEK J. (1985) Stuttering secondary to nervous system damage. In CURLEE R., PERKINS W. (eds.) *Nature and treatment of stuttering: New directions*. London, Taylor and Francis: 31-48.

ROSENBEK J., McNEIL M., PRESCOTT T., ALFREY A. (1975) Speech and language findings in a chronic hemodialysis patient: A case report. *Journal of Speech and Hearing Disorders* 40: 245-252.

ROSENBEK J., MESSERT B., COLLINS M., WERTZ R. (1978) Stuttering following brain damage. *Brain and Language* 6: 82-96.

ROSENFIELD D. (1972) Stuttering and cerebral ischemia. *New England Journal of Medicine* 287, 2: 991.

ROSENFIELD D., MILLER S., FELTOVICH M. (1980) Brain damage causing stuttering. *Transactions of the American Neurological Association* 105: 1-3.

ROTH C., ARONSON A., DAVIS L. (1989) Clinical studies in psychogenic stuttering of adult onset. *Journal of Speech and Hearing Disorders* 54: 634-646.

ROUSEY C., ARJUNAN K., ROUSEY C. (1986) Successful treatment of stuttering following closed head injury. *Journal of Fluency Disorders* 11: 257-261.

RUBENS A. (1975) Aphasia with infarction in the territory of the anterior cerebral artery. *Cortex* 11: 239-250.

RUDMIN F. (1984) Parent's report of stress and articulation oscillation as factors in a preschooler's dysfluencies. *Journal of Fluency Disorders* 9: 85-87.

SAPIR S., ARONSON A. (1990) The relationship between psychopathology and speech and language disorders in neurologic patients. *Journal of Speech and Hearing Disorders* 55: 503-509.

SCHALTENBRAND G. (1975) The effects on speech and language of stereotactical stimulation in thalamus and corpus callosum. *Brain and Language* 2: 70-77.

SCHILLER F. (1947) Aphasia studied in patients with missile wounds. *Journal of Neurology, Neurosurgery, and Psychiatry* 10: 183-197.

SHTREMEL A. (1963) Stutter in the left parietal syndrome (in Russian). *Zhurnal nevropatologii i psikhiatrii imeni S.S. Korsakova* 63, 6: 828-832.

SOROKER N., BAR-ISRAEL Y., SCHECHTER I., SOLZI P. (1990) Stuttering as a manifestation of right-hemisphere subcortical stroke. *European Neurology* 30: 268-270.

SOUQUES A. (1908) Palilalie. *Revue Neurologique* 16: 340-342.

STANSFIELD J. (1995) Word-final dysfluencies in adults with learning difficulties. *Journal of Fluency Disorders* 20: 1-10.

STERLING W. (1924) Palilalie et le symptome 'linguo-salivaire' dans le parkinsonisme encéphalitique. *Revue Neurologique* 35: 205-220.

STEWART T., GRANTHAM C. (1993) A case of acquired stammering: The pattern of recovery. *European Journal of Disorders of Communication* 28: 395-403.

STEWART T., ROWLEY D. (1996) Acquired stammering in Britain. *European Journal of Disorders of Communication* 31: 1-9.

THOMAS P., LALAUX N., VAIVA G., GOUDMAND M. (1994) Dose-dependent stuttering and dystonia in a patient taking clozapine. *American Journal of Psychiatry* 151: 1096.

TIPPETT D., SIEBENS A. (1991) Distinguishing psychogenic from neurogenic dysfluency when neurological and psychological factors coexist. *Journal of Fluency Disorders* 16: 3-12.

TRENEL X., CRINON X. (1912) Palilalie chez une pseudo-bulbaire. *Revue Neurologique* 23: 676-680.

WALLEN V. (1961) Primary stuttering in a 28-year-old adult. *Journal of Speech and Hearing Disorders* 26: 394-395.

WEINER A. (1981) A case of adult onset stuttering. *Journal of Fluency Disorders* 6: 181-185.

M.M. LEAHY, T. STEWART

IDIOPATHIC STUTTERING ONSET IN ADULTS

The voluminous literature that exists on the subject of stuttering attests to the notion that the phenomenon has been perhaps the most widely researched area in speech and language pathology. The result of this is that there is a constant flow of knowledge generated about the problem and about how to work effectively with people who stutter. We can say with cautious optimism as we approach the end of the twentieth century, that several mysterious and controversial issues about stuttering may soon be resolved. Attention in recent studies to the neuropsychology of stuttering, genetic aspects of the disorder, and emerging neurophysiological data opens the possibility of new interpretations of fluency and dysfluency which may also have implications for reinterpreting other speech and language disorders. However, some elements of the unknown about stuttering are likely to remain. The nature of adult-onset stuttering is an area that is currently receiving significant attention, perhaps because of the increasing reports of the incidence of the phenomenon. It would seem as if the majority of clinicians are likely to be involved in the management of adult-onset stuttering in the course of their professional life.

THE ONSET OF STUTTERING

Stuttering is a phenomenon that is generally regarded as having its onset in childhood (Andrews, Craig, Feyer, Hoddinott, Howie, Neilson, 1983; Van Riper, 1982) and when this is the case, the stuttering is regarded to be idiopathic or of cause (or causes) unknown. Adult onset of stuttering has been directly linked to central nervous system damage, particularly when associated with acquired aphasic conditions or acquired motor speech disorders (see Lebrun, this volume). Neurogenic stuttering has also been described as a syndrome that "can exist in its own right" and "not simply a secondary psychological consequence of brain damage" (Bijleveld, Lebrun and van Dongen, 1994:253).

There is increasing evidence however that stuttering, some of which may be of an idiopathic nature, may begin in adulthood (e.g., Stewart and Grantham, 1993; Market, Montague, Buffalo and Drumond, 1990). In some such instances, onset may be linked to recent trauma experienced or to psychological states e.g., depression and anxiety.

In this chapter, issues regarding the nature of what we consider to be non-neurogenic adult onset stuttering, its symptomatology and the development of symptoms over time will be discussed. As there are very

Y. Lebrun (ed.), From the Brain to the Mouth, 139–152.

little documented intervention studies in this area, therapy is by nature experimental. However, some of the therapy ideas that have been used to remediate developmental stuttering have been successfully applied to this client group. Therapies which have facilitated fluency and addressed the problems associated with the symptom will be described. A series of case study data will be presented in conclusion.

SYMPTOMATOLOGY: OVERT AND COVERT SYMPTOMS

It has been suggested that the symptomatology of acquired and developmental stuttering may closely resemble each other (see Lebrun, this volume). As with many other areas in the study of fluency disorders, there are, as yet, few absolutes. However, some preliminary evidence from the few documented examples of adult onset idiopathic stuttering suggest that there may be some factors in the presenting symptoms that may be used in differential diagnosis. Stewart and Grantham (1993) report the occurrence of dysfluency for Client C on approximately 50% of initial consonants, consonant clusters and/or syllables (cv) of words. These dysfluencies were repetitive in nature with between one and six units repeated initially but this level reduced over time as recovery progressed. No struggle or tension were associated with these symptoms at onset but this changed over time with facial and laryngeal tension becoming evident approximately 6 months post onset. Such symptoms are more similar than dissimilar to the more usual course of developmental stuttering.

In other instances, where onset is directly linked to psychological trauma and/or anxiety, struggle and tension appear to be associated with the overt symptoms from the outset. Clinicians report the occurrence of silent and audible prolongations in these clients. In such instances, symptoms may reflect the event which has precipitated the onset of stuttering while in the former cases, the development of struggle and tension may be linked to the stuttering itself as a reaction to it (e.g., from anxiety caused), or as a product of it (i.e. an attempt to change the symptom).

Secondary symptoms: An important finding that emerges from the British survey data (Stuart and Rowley, 1996) is the presence of secondary symptoms that may be associated with late onset stuttering where there is no identifiable cause. Symptoms such as loss of eye contact and tension in the lower jaw and facial muscles in association with the stuttering symptoms, are those that are most frequently reported by clinicians for this group. It is probable that secondary symptoms may be described in a similar way to the occurrence of struggle and tension within speech behaviour. When stuttering may be attributable to a psychological event and/or high levels of anxiety, the secondary characteristics will be present from the onset. In other instances, these same behaviours may result from negative reactions to the stuttering and thus, develop over time.

LEVELS OF SEVERITY

A wide range of severity of symptomatology is reported in instances of adult onset stuttering. There seems to be no obvious correlation between the nature of the acquired stutter and the severity of symptoms in the British study, with idiopathic stuttering presenting as no worse or indeed no better than stuttering acquired from more readily identifiable causes.

COVERT SYMPTOMS

The covert or psychological phenomena which occur with idiopathic stuttering of adult onset frequently present the most difficult to manage aspects of the disorder. When viewed in isolation, such phenomena are complex; when combined with a number of other symptoms and with speech dysfluency, they can present a potent, dehabilitating mixture.

Anxiety: Case studies indicate that anxiety may either precipitate late onset of stuttering or may result from the stuttering itself. Stewart and Rowley (1996) report that anxiety is associated with acquired stuttering in approximately 83% of cases in their survey. Stuttering is reported to occur in anticipation of a major traumatic event (e.g., a court appearance, an important interview) and there are other examples where the anxiety is associated with recall of a past event (e.g., a bomb blast). Since anxiety can be generated from either anticipation of a reoccurrence or recall of the traumatic incident itself, it can be linked at any point in time to the trauma. Such anxiety is compounded by anxiety resulting from the stuttering event. This seems to follow patterns of anxiety associated with stuttering, which may be linked with the negative stereotype of the stutterer, or from construing of stuttering as 'bad'. Once stuttering is experienced as loss of control, it can become all-encompassing and clients may begin to think negatively about themselves in terms of their ability generally, and also perhaps in terms of self-consciousness, unpleasantness, nervousness, tension, reticence, fearfulness, unfriendliness, and introversion (Leahy, 1994). This aspect of the problem is further discussed below.

Grief (or *Kellian guilt*): Processes akin to the experience of grief or loss have been observed in individuals with late onset of stuttering. Initially, some evidence of denial has been observed, not necessarily of the stuttering itself but of the impact of the stuttering on the client's life, for example, the client believes that the change enforced by the stuttering will not necessitate modification of any other aspect of behaviour. Generally, such a coping mechanism is short-lived but it can return from time to time. One interpretation is that the individual is clinging to the raft of familiarity in an attempt to make some sense of a changing sea.

Sadness or grief for the 'loss of self' can occur too as a reaction to late onset of stuttering. This differs from the perceived loss reported by other adults who have stuttered since childhood, e.g., loss of childhood, loss of career, unfulfilled relationships. These "new stutterers" who previously considered themselves to be good communicators, spontaneous, social, fluent and informative, can begin to believe that an important and perhaps core aspect of themselves has been lost. Essentially, strong and basic aspects of identity are drawn into play and the threat associated with a major change in status may be overwhelming. Kelly's (1955) constructivist idea of *guilt* is experienced. This is similar in one sense (but opposite in another sense) to the kind of situation that occurs when a person who has stuttered since childhood — and for whom the stuttering plays an important role in personal and social terms — finds that by becoming fluent, he may lose what is part of his core role structure, an event that has greater implications than the achievement of fluency itself (Hayhow and Levy, 1989). Stewart and Grantham (1993) report that Client C commented upon feelings of Kellian guilt on a number of occasions: "I suppose I was mourning the loss of my 'old self', the person I was, because in order to cope I've had to change. I feared most of all a loss of identity, that C. would become 'the one who stammers' ". Likewise, Client D (see below) described how stuttering became "the most important aspect of my life, something I constantly analysed and tried to both 'cure' and accept as part of my identity, but could do neither".

Clients frequently have to confront their loss over and over again. Initially this can be difficult and painful and is something which individuals may choose to avoid. Clients have been observed to withdraw completely at some point, not wishing to go out or talk with others so that their stuttering and the sense of loss of self will not be exposed. Feelings of vulnerability are too strong to risk being revealed to those outside. At a later stage, when it becomes clear that some risk-taking is appropriate, the person who stutters often finds that some explanation is needed when talking to friends and acquaintances. Often, this requires dealing with their stereotypes and consequently, feelings of shame may be experienced. When meeting new acquaintances, some clients choose to discuss the onset of their stutter in an attempt to explain it away e.g., in terms of "this is not really me, it is just something that has happened to me".

Reflection on the possible cause of their newly acquired stuttering is a common aspect of the process of coming to terms with it. Absence of neurological findings may be a source of further confusion for some clients and there may even be rejection of psychological explanations or hypotheses regarding causative factors. In some instances, there is a tendency to begin with 'self' as the source of blame, a situation which can be counter-productive and disheartening.

REACTIONS AND RELATIONSHIPS

Fear and vulnerability are likely to be experienced following the onset of stuttering. Clients are particularly susceptible to adverse listener reactions or may perceive some listeners to be reacting negatively, when in some cases, they are merely curious or puzzled by this new stuttered speech. Family and close friends can be unsure how to help the individual's speech and emotional state. Unrealistic expectations from family and friends about recovery and therapy outcome are not uncommon. Involvement of family and/or friends in the therapy process can be extremely useful to help reduce undue pressure on the client and to provide a supportive environment where the client's personal needs are respected. However, clients may not wish to involve members of their family in the therapy process as this may seem to them to compromise their independence and sense of responsibility, and possibly compound some feelings of helplessness, at least at the initial stages of therapy. This contrasts with one instance of recently acquired stuttering (see Client E below) in a young man of 18 years who had been on medication to control epilepsy for 6 years. Three members of Client E's family attended the initial sessions and all three took an active part in therapy process, which proved to be successful within the relatively short timespan of c. 9 weeks.

Experience of the negative 'stutterer' stereotype: Adults who begin to stutter will probably have an awareness of the commonly held negative stereotype of stutterers and therefore they are likely to experience their stuttering as negative from the start. This may be an important difference between developmental stuttering and adult-onset stuttering, as the sense of loss of the (fluent) self and the grief experienced as a result are likely to be deeply and sharply felt. Sense of self in the child is in the process of being acquired and is not dependent upon fluency as it may be in some adults.

People who stutter are generally categorised as belonging to an undesirable group, one where there is every possibility of being judged as 'deviant' and inferior in many ways. The influence of the stereotype is relatively pervasive and the person who stutters is likely to experience social rejection generally (Van Riper, 1982) and prejudice in the job market (Hurst and Cooper, 1983; Silverman and Paynter, 1990). The form of stuttering behaviour exhibited is probably influenced by the speaker's awareness of the stereotype and knowledge of the possible reactions of listeners (Starkweather, 1987). Johnson (1946) described "the uncanny, crushing power of the social disapproval of whatever is regarded as stuttering as frightening, perplexing and demoralising" (reported in Van Riper, 1982:226). Evidence exists to show that the person who stutters may experience feelings of inadequacy, powerlessness, helplessness, hopelessness and failure, as well as negative self-concept, guilt and shame (Perkins, 1992; Van Riper, 1982). Such feelings probably hold as true for many of those adults who have recently begun to stutter as for those who have a longer history of stuttering. Client C reported feelings that ranged from annoyance to anger when people reacted negatively to her stuttering:

"Other people annoyed me because so many of them reacted badly to stammerers. Most people at one time or another either interrupted or finished off my sentences. This was very, very annoying, particularly as they often got the word or sentence wrong. They thought they were helping but in fact they were doing the complete opposite. Those people were fairly easy to deal with. Harder to cope with were the ones I might meet only once but who treated me like a child. Since beginning to stammer, people — especially in shops — began smiling at me in a very condescending manner as soon as I said anything. I found attitudes like this very hard to cope with. I felt insulted and degraded. The last group of people I have discovered would rather avoid or ignore anyone who does not sound 'normal' rather than speak to them. I gradually came to the conclusion that it was because they did not know how else to react. They were suddenly faced with a situation they have never come across before and which they did not understand".

PROGRESSION AND CHANGE OF SYMPTOMS OVER TIME

The final issue regarding symptomatology to be discussed here is the progression and change of symptoms over time. Whether the patterns noted in adult-onset stuttering bear any similarity or relationship to those of childhood stuttering is a question that cannot be fully addressed here as the available number of case studies is limited. However, some initial findings are worthy of note.

Stewart and Grantham (1993) charted the progress of Client C over a 12 month period and observed a marked increase in fluency at approximately 4 months post onset. At this point, the client's overt symptoms of repetitions decreased by c. 50% with an accompanying increase in rate of speech. As time went on, it was also noted that the client developed a number of secondary speech characteristics such as struggle and avoidance. These seemed to occur independently of any increase in fluency and required constant monitoring by the therapist.

In contrast to this, Client D, a 27 year old lecturer (whose onset of stuttering occurred at age 22 years), reported that before beginning therapy, her stuttering gained a level of constancy in the teaching situation that only varied to any significant degree during holiday time when she was not under pressure to perform. This pattern continued during the course of therapy, with fluency being achieved following a break of some weeks from teaching. Although the stuttering varied in frequency according to the anxiety generated in anticipation of particular situations, the pattern of stuttering did not change to any significant degree. Secondary symptoms of struggle did not emerge but avoidance of some words and situations did occur but these were dealt with effectively (see case study data below).

ASSESSMENT: SUGGESTED PROTOCOL

The assessment of the client with recent onset of stuttering should begin with a detailed case history which will focus on the events that have precipitated the stuttering. It is recommended that an analysis of the overt symptoms take place but the therapist should be sensitive to the fact that many clients experience stress when confronting the stutter. A videotape recording should therefore be used with care. It is useful to monitor the progression of symptoms over time and this should be recorded to contribute to the available data on the disorder.

Overt symptoms: Abnormal breathing patterns have been noted in many clients presenting with adult-onset stuttering. Thorough investigation of breathing to include a focus on lung capacity, control and focus of breath management and overall adequacy is recommended.

Facilitating the client in analysis of the stuttering and fluency is more important in our view than elaborate objective estimates of frequency of stuttering. Description of types of dysfluencies in terms of overt and covert symptomatology is vital.

Covert symptoms: Avoidance at all levels, especially word and situation avoidance should be discussed. It is recommended that some objective measure of anxiety be administered, e.g., the Hospital Anxiety and Depression Scale (HAD Scale, Zigmond and Snaith, 1983) would be useful here as it identifies the level of anxiety experienced, denotes evidence of clinical depression and can chart progress/reduction in symptoms.

Struggle and tension: In order to prevent the development of secondary symptoms, it is essential that careful attention is paid to these elements throughout the therapy process. Client-based scaling techniques are useful here (e.g., the client is asked: "*On a scale of 1 to 10, how tense do you feel today*?"). The site of the tension or struggle should also be noted because of fluctuations in behaviour. For some clients there is a link between breathing and the feeling of tension in the upper respiratory system.

Focus on variability of symptoms may help both therapist and client understand the patterns of behaviour associated with the stuttering and perhaps this may open the avenue to hypothesising about the nature of the problem. In the case of Client D, issues that emerged from such hypotheses included the linking of stuttering with child-like characteristics, whereas fluency was associated with more adult characteristics. Stuttering has also been associated with stress and extreme anxiety when variability was looked at in some detail. Repertory grids (Fransella and Bannister, 1977; Hayhow and Levy, 1989) are helpful and powerful in the analysis of construing both role relationships and anticipation of situations which shed light on the variability of stuttering symptoms.

APPROACHES TO THERAPY

Adult clients presenting with recent onset of stuttering will be offered thorough medical and neurological examination in order to eliminate or reduce the possibility of associated pathology. In many ways, therapy may be considered to be experimental as the usual approaches to working with stuttering of long standing may not be appropriate. Some approaches to working with clients considered to have acquired idiopathic stuttering as adults that have proved useful will be outlined here.

Discussion of possible causes and/or theoretical explanations of stuttering is an important part of the therapist's brief during the early stages of therapy. The fact that developmental stuttering is of unknown causation should be clearly stated in order to help put the present problem into context. Knowledge about the usual course of therapy and indications of how stuttering may be successfully managed, if not eliminated, should be presented. Issues of concern to the client have to be addressed in as much detail as may be required so that a clear understanding between client and therapist emerges. Because of the elements of mysterious causation and the implicit psychological nature of the stutter, forging a good working relationship will be enormously helpful in addressing facilitation of fluency control.

Direct focus on fluency: Learning fluency techniques may be an important step for some clients even when the stuttering is of relatively recent origin. This may help open the possibility of a speedy resolution of the problem, as was the case with Client E (see case study below). Learning about the control of breathing and the elements of fluency proved to eliminate the severe symptoms of laryngeal blocking and tense, struggle reactions that had begun. By providing early intervention in this instance, the client along with his family felt able to manage the symptoms and did so in a relatively short time. Negative feelings associated with the stutter did not develop.

An open approach to therapy is necessary as there are no tried and tested techniques specific to this client group documented. We recommend that therapists present clients with an armoury of approaches from which together, they select one or two techniques for experimentation. Should these be less useful, client and therapist can revisit the armoury and check out a different set of strategies. Experimentation may include trying different ways of being fluent which allows the client to experience levels of control. This can facilitate a new understanding of stuttering and fluency, e.g., that stuttering is but a different way of speaking and that there may be several different ways of managing it. This loosening of construing invites the possibility of change back to fluency and may be extremely encouraging for some clients.

Elaborating the meaning of the stutter may be achieved in a variety of ways. Different media can be exploited and matched to the client's style, e.g., Client C responded well when asked to draw a representation of her stutter. The subsequent illustration was useful throughout the therapy process and as a focus for C's anger.

Focus on covert symptoms: Because of the perplexing and debilitating effects of adult-onset stuttering, working to facilitate the reduction of anxiety and the management of stress associated with the stutter could be crucial aspects of therapy for some clients.

Anxiety: Where the onset of stuttering is related by the client to a particular traumatic event, there may be a fear of recurring similar events that can cause much anxiety. This may represent a feeling of being caught in a dilemma without the possibility of escape or with the feeling that one is not able to cope. Feelings of helplessness or of fear may be approached in several different ways, e.g., it can be useful for the client to explore different ways of construing similar situations and in turn, the actual event or circumstances that are feared. The exploration of the event from a range of different perspectives with the possibility raised later of seeing one's own perspective as more similar (or less different) to that of others, and of reducing some of the more negative or feared attributes of the event. Other options include: enactment of the events as the client might like them to occur; envisaging oneself as a different participant, or envisaging the scene as better or worse than it actually was (see also: Kelly's (1955) fixed role therapy).

The control of anxiety may also be managed with techniques such as that proposed by Snaith (1981) in Anxiety Control Therapy (ACT). This approach has been successfully exploited with adults who stutter both as a practical means of controlling high levels of anxiety experienced before and during speaking situations, and perhaps more importantly, as a PERCEIVED, tangible belief that they can manage these feelings which have previously seemed out of their control (Turnbull, 1987).

Stress management may be approached by considering those activities that the client enjoys that do not depend on speaking and scheduling time for them in the daily routine. Appreciating the importance of relaxation and recreation, along with the importance of oneself and giving time to oneself, facilitates the management of pressure so that stress may be reduced. Implications of taking control of aspects of time management may have effects on working with the stuttering symptoms.

Grief (or '*guilt*'): When an adult who has begun to stutter expresses a sense of loss of identity or of role, it is often necessary to focus on the person, as opposed to the speaker (or stutterer) — as a means of instigating recovery. For example, with Client C, the healing process began by the therapist's statement: "You stammer, so what? You talk slightly differently, that doesn't change the person you are". Character sketches looking at the differences between the client's construing of self before the stuttering began and now, and of the 'ideal me' may help with this process and may also provide insightful information for therapy focus (see also Hayhow and Levy, 1989). Affirmation of the person and direct attention to the means of construing oneself positively will help alleviate aspects of the stress associated with stuttering and will support the momentum of change.

CONCLUSION

Achieving desired change or return to a previous status is not always an easy process. Having helped specify aspects of the stuttering and associated problems, the therapist's role includes listening carefully and providing support and encouragement for the client. In addition to direct work with the client, the therapist should also be concerned with the support systems in the immediate environment of the client. Advice should be given to family members and significant others on how to help the client, dealing positively with the stutter, keeping the stutter 'on the agenda' and providing support and comfort.

Challenging the client from time to time should be discussed, indicating when risks are necessary for change and when they might be best faced another time. Monitoring changes and facilitating client self-reliance and self-monitoring are part of the means of the client becoming one's own therapist, the ultimate aim of therapy.

It is clear from the approaches taken to understanding and managing the adult-onset idiopathic stuttering as described in this chapter, that our impression is that the symptoms presented are probably psychogenic in nature. Some reservations remain about the nature of the symptoms, and some similarities may exist between this client group, those with post-traumatic stress disorder and possibly those with spasmodic dysphonia. Nevertheless, we are optimistic that recovery from adult-onset stuttering can be facilitated, using approaches such as those outlined in this chapter

There follows a selection of case study data that demonstrate a range of differences in presentations of adult-onset stuttering.

Client A: Female.
Personality: A was described as a bubbly, chatty person who was also moody and impatient. She appeared to be confident but often this hid underlying anxiety.
Family history: A's parents divorced when she was young. Her mother re-married and she saw her father at weekends. She had two older step sisters.
Medical/social history: A suffered from eczema between the ages of 5 and 11 years. She had difficulty articulating /s/ as a child. From puberty, A is reported as being overweight and when older and a student at music college, she sought psychiatric help for her weight problem and for the stress headaches she experienced. There were no neurological symptoms at time of onset.
Onset of stuttering: A began stuttering at age 26 years. At the time of onset

she reported feeling unhappy at work. She was involved in a long-standing relationship and her partner had just recently lost his job. Her dysfluency started during a weekend visit to her parent's home when she was accompanied by her partner for the first time. She noticed the stutter herself and it was also remarked upon by other members of the family.
Symptoms: Silent blocks and loss of eye contact were the primary symptoms. In addition, A reported feeling anxious, nervous and frustrated.
Therapy: A attended for therapy which concentrated on attitude work and counselling.
Outcome: No data available.

Client B: Female.
Personality: B was described as outgoing, lively and caring.
Family history: B's mother had a CVA some years before the onset of B's stutter.
Medical/social history: B suffered from hypertension for 30 years and anxiety and depression for 10 years prior to starting to stutter. There were no neurological symptoms at time of onset.
Onset: B (aged 35 years) was due to appear in court shortly before her stutter began. She reported feeling very anxious about this and was not sleeping well.
Symptoms: Repetitions and silent prolongations (blocks) were present in B's speech. Syntax was telegrammatic on occasions. An adaptation effect was noted on reading the same passage three times. She also had reduced eye contact and word avoidances. B reported feeling fear and anger and had noticed that the stutter seemed to occur when communication was most important.
Therapy: B attended for therapy and was taught easy onset and some block modification techniques. Attitude to speech was also a focus in therapy and counselling was a part of the therapy process.
Outcome: Following a period of therapy B's stutter was greatly reduced with the introduction of controlling techniques but elements of the stutter remained. B's attitude was considered to be more positive, despite a difficult domestic situation continuing.

Client C: see Stewart and Grantham (1993).

Client D: Female.
Personality: D is bright, extroverted and generally considered to be a happy, social person with many friends.
Family history: D is the oldest of a family of 5 children. Her uncle and his

daughter have a history of stuttering. No other significant family history.
Medical/social history: No significant medical history, no physical problems at time of onset of stutter. Previously experienced dietary problems that were regulated without medication. Beta-blockers are taken occasionally when D is faced with a particularly stressful event. Absence of neurological symptoms.
Onset: D reported that stuttering began at age 22 years while a final year student at College. The stutter became progressively worse during 3 years of post-graduate PhD research work and reached 'crisis point' when faced with lecturing to very large groups (300+), aged 27 years. *Symptoms*: Rapid, effortless, 3/4 unit syllabic repetitions word-initially are the predominant feature of D's stutter. Tension is reported but was evident only on three occasions in reading, when momentary blocking occurred. Frequency of repetitions increases in pressure situations. Speech is generally stutter-free when relaxed and out of term time.
Therapy: D had attended group therapy sessions for a period of 7 months before taking up the post as lecturer. Stress reduction, breathing for relaxation and control, and fluency techniques were all part of therapy but the major focus was on understanding the variability of stuttering, building confidence and feeling in control of situations.
Outcome: Following therapy over an intermittent period of c. 14 months, D continues to stutter when lecturing but is sometimes able to control fluency to some extent. When last contacted, was considering different, less stressful career options.

Client E: Male.
Personality: E is described as 'quiet but social' and very popular with friends of both sexes his own age. *Medical/social history*: At age 12 years, E suffered cerebral haemorrhage as result of rugby accident. Developed epilepsy soon after this and was taking medication (Tegretol and Benntin) for 6 years before severe stuttering began. Mother reported that some dysfluencies were evident infrequently in E's speech but considered these to be attributable to word-finding difficulties. Neurologist considered that the stutter could be associated with medication but doubted if this were so.
Onset: E began to block severely at age 18 years.
Symptoms: Blocks lasted up to 6/8 seconds on plosive sounds and were accompanied by extraneous movements of limbs. Frequency was up to 50% of words.
Therapy: Direct focus on block modification and breathing control. Three members of family attended for direction and to facilitate E's therapy.
Outcome: E's acquired fluency at a steady pace in a relatively short pe-

riod (c. 9 weeks), without attending for regular therapy. His stutter had completely resolved on follow-up.

REFERENCES

ANDREWS, G., CRAIG, A., FEYER, A., HODDINOTT, S., HOWIE, P., NEILSON, M. (1983). Stuttering: A review of research findings and theories circa 1982. *Journal of Speech and Hearing Disorders*, 48, 226-246.

BIJLEVELD, H., LEBRUN, Y., van DONGEN, H. (1994). A case of acquired stuttering. *Folia Phoniatrica et Logopaedica*, 46, 250-254.

FRANSELLA, F., BANNISTER D. (1977). *A manual for repertory grid technique*. New York: Academic Press.

HAYHOW, R., LEVY, C., (1989). *Working with stuttering*. Bicester: Winslow Press.

HURST, M., COOPER, E. (1983). Employer attitudes toward stuttering. *Journal of Fluency Disorders*, 8, 1-12.

JOHNSON, W. (1946, in C. Van RIPER, (1982), *The nature of stuttering*, 2nd Edition. Englewood-Cliffs, NJ: Prentice-Hall.

KELLY, G. (1955). *The psychology of personal constructs*. New York: Norton.

LEAHY, M. (1994). Attempting to ameliorate student therapists' negative stereotype of the stutterer. *European Journal of Disorders of Communication*, 29, 39-49.

MARKET, K., MONTAGUE, J., BUFFALO, M., DRUMOND, S. (1990). Acquired stuttering: Descriptive data and treatment outcome. *Journal of Fluency Disorders*, 15, 21-33.

PERKINS, W. (1992). *The prevention of stuttering*. London: Whurr.

SILVERMAN, F., PAYNTER, K. (1990). Impact of stuttering on perception of occupational competence. *Journal of Fluency Disorders*, 15, 87-93.

SNAITH, R. (1981). *Clinical neurosis*. Oxford: Oxford University Press.

STARKWEATHER, W. (1987). *Fluency and stuttering*. Englewood Cliffs, NJ: Prentice-Hall.

STEWART, T., GRANTHAM, C. (1993). A case of acquired stammering: the pattern of recovery. *European Journal of Disorders of Communication*, 28, 395-403.

STEWART, T., ROWLEY, D. (1996). Acquired stammering in Britain. *European Journal of Disorders of Communication*, 31, 1-9.

TURNBULL, J. (1987). Anxiety control training and its place in stuttering therapy, in C. Levy (Ed.) *Stuttering therapies: Practical approaches*. London: Croom Helm.

Van RIPER, C. (1982). *The nature of stuttering*. Englewood NJ: Prentice Hall.

ZIGMOND, A., SNAITH, R. (1983). The HAD Scale. *Acta Psychiatrica Scandinavica*, 67, 361-370.

J. VAN BORSEL, K. VAN LIERDE, K. OOSTRA, C. EECKHAUT

THE DIFFERENTIAL DIAGNOSIS OF LATE-ONSET STUTTERING

INTRODUCTION

Neurogenic stuttering, by definition, is stuttering caused by damage to the central nervous system. In this respect neurogenic stuttering clearly differs from developmental stuttering of which the cause to date is unknown. The fact that in neurogenic stuttering there is an evident cause does not mean, however, that the diagnosis of neurogenic stuttering is always straightforward. As reported in a number of studies, stuttering, when not developmental, may also occur as a side-effect of medication (see for instance Nurnberg and Greenwald, 1981 or McClean and McLean, 1985). Also, cases of non-developmental stuttering have been reported in which the stutter was of psychogenic origin (see for instance Deal, 1982; Deal and Doro, 1987; Roth et al., 1989; Tippet and Siebens, 1991; Mahr and Leith, 1992). In this paper a case of stuttering is presented which illustrates this problem of differential diagnosis in non-developmental stuttering. The patient presented with stuttering for the first time after she had suffered brain damage. Yet, there are reasons to believe that her dysfluency was not of neurogenic origin.

CASE HISTORY

I.V., a 15-year-old girl, was admitted to the University hospital after she had been overthrown by a motor vehicle. She had a cerebral trauma, a fracture of the acetabulum, a fracture of the pubis and a hemopneumothorax. Admittance CT of the brain showed a small intracerebral bleeding near the left cerebral peduncle. A control CT performed a few days later revealed also a hyperdense area at the crus posterior of the internal capsule on the left with surrounding oedema. On admission to the hospital I.V. was comatose and remained so for five weeks. Her condition gradually improved and four months after the accident she was transferred from the neurosurgery ward to the hospital rehabilitation center. Neurological examination at that time revealed that sensibility was not impaired but that there was a spastic hypertonia on the right hemisoma.

Neurolinguistic screening about 5 months after the accident showed that comprehension of both oral and written language was fairly good. I.V. was

Y. Lebrun (ed.), From the Brain to the Mouth, 153–161.

perfectly able to execute simple and semi complex commands whether presented orally or in written form. Her score on the Dutch version of the Token Test (Van Dongen et al., 1976) was reduced though (49/61). Also oral expression appeared to be virtually impossible. I.V. could not produce any audible sound, either voiceless or voiced. Respiratory control was moreover very poor and breathing could not be prolonged for more than a few seconds. Volitional oral movements were either slow (to wiggle one's tongue, to lick one's lips, to pucker one's lips, to show one's teeth) or impossible (to click one's tongue, to kiss). Neither did I.V. manage to hum, clear her throat or cough on request. Written expression on the other hand was quite possible by means of a letterboard. On several occasions during the testing I.V. produced in this mode complete and correctly spelled sentences in reply to questions. On the basis of the above findings it was concluded that I.V.was primarily dysarthric.

At the hospital rehabilitation center I.V. was enrolled for occupational therapy, physical therapy and speech therapy. Individual speech therapy was given daily and focused on improvement of respiratory control, on the elicitation of phonation and on improvement of articulation.

More than half a year after the first examination I.V. was again referred to us. She was still receiving speech therapy. However, part of the individual therapy had been replaced now by group therapy. It was noted that on the one hand I.V. was not very motivated any more. Several times she did not turn up for therapy. On the other hand she complained about not receiving more individual therapy. Also her discharge from hospital was pending. There was some discussion, however, whether it would not be more beneficial to delay discharge until after an operation which I.V. still had to undergo to improve mobility. As far as her communication problems was concerned I.V. had made considerable progress. She was no longer dependent on a letterboard to communicate. Her voice was still weak but clearly audible and articulation was fairly slow but otherwise correct. Also her score on the Token Test (56/61) now definitely ruled out aphasia. Moreover on a test of sentence comprehension from a Dutch standardized aphasia battery (the S.A.N., Deelman et al., 1981) I.V. obtained the maximum score. Nonetheless, verbal fluency, as measured by a subtest from the same aphasia battery, appeared to be reduced (7th percentile).The reason for referral this time, however, had not to do with the communication problems discussed so far. I.V. was referred now because from one day to the next she had developped a marked dysfluency.

An analysis of the dysfluency based on video-recordings revealed that I.V. presented only two symptoms. The main type of dysfluency demonstrated were blocks. In a sample of conversational speech with an overall dysfluency index of 24 per 100 syllables spoken, blocks accounted for 81% of the dysfluencies. The remainder of the dysfluencies (19%) consisted of prolongations. Further analysis showed that the blocks and prolongations always occurred on consonants. Vowel sounds were never affected. As a rule blocks and prolongations occurred on syllable-initial consonants. In polysyllabic words it was moreover most often the initial syllable of the word that was affected. Examples of syllable-final dysfluencies were not observed. The blocks and prolongations occurred mainly but not exclusively on lexical words. I.V. produced the blocks and prolongations without any accompanying accessory features such as facial grimacing, movements of the head, blinking of the eyes or quivering of the nostrils. Duration of the prolongations was usually from one to two seconds. Blocks tended to last longer, sometimes up to five seconds. With blocks on alveolar plosives the inferior surface of the tongue would often protrude visually between the incisors while the tip of the tongue made a firm contact with the alveolar ridge and upper incisors.

In addition to conversational speech, samples were recorded and analysed of serial and imitative speech, reading aloud and singing. I.V. appeared to be highly dysfluent in each of these tasks too.

Recitation of the days of the week, the months of the year and the numbers from one to twenty could not be completed without omitting some items which I.V. after several attempts did not manage to utter. Overall this task gave rise to a total of 26 stutters. Just as in conversational speech the majority (18/26, i.e. 69%) of the dysfluencies were blocks. It was during the recitation of automatized series (and only during this task) that we also observed a strange kind of "secondary" behaviour. A few times when I.V. did not manage to utter a word because of a block she resorted to the peculiar strategy of pinching the nose to overcome the block. And indeed this strategy seemed to work and enabled I.V. a few times to unseal a block. Later we heard that it was a therapeutic strategy which I.V. had originally learned to apply to reduce hypernasality at a stage when velopharyngeal closure was still inadequate. One may conjecture that the strategy was successful in unsealing blocks because it temporarily caused a complete closure of the expiratory airway. Since conceivably the voluntary manual closure at the nasal airway was stronger than the unvoluntary oral closure, it was the latter one which of necessity had to be broken up to continue the expiratory airflow.

Blocks and prolongations were also observed during administration of the subtest "Naspreken" from the Dutch version of the Aachener Aphasie

Test (Graetz, De Bleser and Willmes, 1992), a test requiring in succession the repetition of isolated sounds, isolated words and sentences. Just as in the recitation of automatized series, several items had to be omitted which I.V., because of a block, did not manage to utter. And again blocks appeared to be the predominating symptom. Of the total number of 49 stutters displayed during this task 41 (i.e. 84%) were blocks.

The same trend also characterized the dysfluency displayed in an attempt to sing a few lines from a popular Dutch song. I.V. was not able to really sing. Rather she more like recited the words of the song without much prosodic variation. Of the nineteen syllables produced in this way four showed a block and one had a prolongation.

Finally fluency was assessed when reading the Dutch version of the IPA-text "the north wind and the sun were disputing which was the stronger . . ." (International Phonetic Association, 1974). On three successive attempts to read the first two lines of this text (49 syllables) I.V. produced respectively 11 blocks, 9 blocks and 2 prolongations, and 6 blocks and 2 prolongations.

A second evaluation of dysfluency took place about one month and a half after the first one. I.V. was still an inpatient at the hospital rehabilitation center but she was about to be discharged. Using the same series of tests, the picture that was observed now, was not much different from the one that had been observed on the previous assessment. I.V. was still highly dysfluent demonstrating blocks and to a lesser degree also prolongations in each of the conditions sampled. Interdental protrusion of the inferior part of the tongue on blocks on alveolar plosives was still a prominent feature. Yet, it was also noted that I.V. now sometimes avoided stuttering by whispering or by making light consonant contacts. As far as we could ascertain, she must have developed this behaviour by herself. At any rate, none of the speech therapists at the hospital rehabilitation center remembered ever having introduced this strategy. They had moreover never focused therapy directly on eliminating or modifying I.V.'s stutter behaviour. As to attempts to overcome blocks by manual closure of the nostrils, these were no longer seen. In addition it was further assessed during this second evaluation that I.V. did not become more fluent when speaking under masking noise (85 dB masking noise generated by an audiometer, presented binaurally).

Following the evaluation we had a short discussion with I.V. and her mother, who accompanied her, on I.V.'s speech behaviour. I.V.'s mother confirmed that her daughter had developed the dysfluency from one day to the next. And she was sure that I.V. had never stuttered as a child and that there was no family history of stuttering. We assured I.V. that her speech would become better again and insisted that she should not bother about her stuttering. Also we encouraged the use of light consonant contacts.

Three months later we saw I.V. for a third assessment. She was no longer at the hospital rehabilitation center then but attended a school for special education. She looked quite well, actually strinkingly better than during the previous session, and she was clearly very pleased with her new situation. As to her dysfluency a dramatic change was observed. Analysis of a sample of conversational speech revealed not a single dysfluency. Neither did we observe dysfluencies during an attempt to sing, which I.V. still did in a recititative way. The production of automatized sequences (days of the week, months of the year, numbers from one to twenty) could now be achieved without omitting a single item, and only one block was noted. During the repetition task only four minor blocks were observed and I.V.'s reading of a 130 syllable passage of the IPA text *The north wind and the sun*. contained but two blocks. The blocks that were occasionally produced were always of short duration, and visual protrusion of the inferior part of the tongue with blocks on alveolar plosives no longer occurred.

I.V.'s mother told us that the marked improvement had appeared soon after the previous evaluation. She believed that the short discussion we had then with I.V. on her stuttering had triggered the improvement.

DISCUSSION

Developmental stuttering usually has its onset between the ages of two and five years (Silverman, 1992, p. 90). When stuttering occurs for the first time after the typical childhood period several hypotheses should be considered.

One possibility is the one suggested by Van Riper (1971) who warned that in cases of late-onset stuttering there is always some suspicion that the stuttering is a recurrence of childhood stuttering (p. 62). In the present case such an explanation is not very likely. I.V.'s mother denied any prior history of stuttering in her daughter. Neither was there a family history of stuttering as is often seen in developmental stutterers (see for instance Bloodstein, 1995, p. 121, Table 7).

Another possibility which should be considered in cases of stuttering of late onset is that the dysfluency is drug-induced. A number of such cases have been published. Quader (1977) described two patients who developed stutter-like speech following the administration of amitriptyline, an antidepressant. And also in a case described by Elliott and Thomas (1985) stuttering was associated with the administration of an antidepressant (alprazolam). Nurnberg and Greenwald (1981) reported stuttering in two patients as a side-effect of phenotiazine, a neuroleptic. A few authors also reported the occurrence of stuttering consequent upon the administration of phenytoin, an anticonvulsant (Helm, Butler and Canter, 1980; McClean and McLean, 1985). That the stuttering should be drugrelated in the present case is again improbable. During her 11 months' stay at the hospital rehabilitation center I.V. was not administered any antidepressant or anticonvulsant medication nor any neuroleptic drug.

In many cases stuttering which presents for the first time after childhood is of neurogenic origin. Neurogenic stuttering is by far not as common a condition as developmental stuttering, but still a substantial number of cases of stuttering following brain damage have been described by now (for reviews see Rosenbek, 1984; HelmEstabrooks, 1986, 1993 and the present volume). It has become clear that neurogenic stuttering is not associated with any particular lesion. It may occur in conjunction with damage to the low or high brain stem, to the basal ganglia, left and right cerebral hemispheres, and to the white matter of the frontal lobes of either the right and left hemisphere. It has also been reported after frontal, parietal, and temporal lesions within the left hemisphere. (Rosenbek, 1984). Since I.V. became dysfluent after a motor vehicle accident resulting in documented brain damage, the hypothesis that her stuttering was of neurogenic origin would seem very plausible. Moreover her dysfluency showed several of the characteristics which, according to some authors at least (see for instance Helm, Butler and Benson, 1978, Helm, 1993), are typical of neurogenic stuttering. Among these are the observation that the stuttering occurred across speech tasks (I.V. also produced dysfluencies in serial speech tasks and during attemps to sing), that speaking under loud masking noise did not result in more fluent speech, that the stuttering was unaccompanied by accessory behaviour such as facial grimacing or eye blinking, that dysfluencies occurred on both function words and high-information words and that there was no clear evidence of an adaptation effect in repeated reading of a passage. In addition the nature of I.V.'s dysfluencies was such that the possibility of it being another type of dysfluency of neurogenic origin, palilalia, can be ruled out. In palilalia, which generally results from bilateral brain damage, there is typically a compulsive repetition of speech elements (usually longer elements, such as words, phrases and sentences). I.V. did not show any compulsive repetition of speech elements. Blocks, which were the predominant feature of I.V.'s dysfluency, are not usually seen in palilalic patients and neither are prolongations, the only other dysfluency she displayed next to blocks.

And yet, there is some doubt whether the dysfluency observed in I.V. was definitely neurogenic stuttering. At least some of the features which are considered indicative of neurogenic stuttering by some investigators, have also been reported by others to be typical of psychogenic stuttering, i.e. stuttering as a reaction to psychological disturbances. According to Deal (1982), for instance, symptoms such as the occurrence of the dysfluency across different speech conditions and the absence of secondary features are also typical of psychogenic stuttering. Other traits that are considered characteristic of psychogenic stuttering are a sudden onset and coincidence with psychological pressure. I.V.'s mother reported that her daughter began to stutter from one day to the next. And her dysfluency appeared more than a year after the accident, in a period when she was no longer motivated for therapy, while on the other hand there was some uncertainty about her discharge from hospital. In neurogenic stuttering the occurrence of dysfluency is usually more directly associated with the neurological insult. Data from Market et al. (1990) suggest that the large majority of neurogenic stutterers (more than 80%) start to stutter within one month of their insult. Still another feature listed among the characteristics of psychogenic dysfluency by Tippet and Siebens (1991) is that a decrease in the dysfluency is usually accompanied by a general improvement in the patient's outlook, attitude and self-concept. Interestingly, when we saw I.V. for the third time she looked strikingly well and was pleased with her new situation.

The strongest suggestion for psychogenicity in I.V.'s case is probably the pattern of recovery. Neurogenic stuttering is often reported to persist for a long time and often seems to have a poor response to treatment.(see for instance Andrews, Quinn and Sorby, 1972; Helm, Butler and Benson, 1978; Fleet and Heilman, 1985; Ardila and Lopez, 1986; Lebrun, Bijleveld and Rousseau, 1990; Stewart and Grantham, 1993) By contrast, reversibility of the symptoms is considered one of the main characteristics of psychogenic stuttering (Tippet and Siebens, 1991). I.V.'s speech improved quickly soon after the second evaluation. Her mother related the improvement to the discussion we had at that time with I.V. on her stuttering. What we did was to state that her speech would become better and to encourage her to use light consonant contacts as substitutes for blocks. Optimism on the part of the therapist and a symptomatic approach have been advocated and found useful by other investigators too (see Tippet and Siebens, 1991) in cases of psychogenic stuttering. Whether it was really the discussion we had with I.V. that was responsible for the improvement is, however, hard to say for certain. It was also at the same time that I.V. was finally discharged from hospital and that she started special education. Perhaps her improvement was also related, then, to this change of environment that had been pending for several weeks.

A final hypothesis that should be considered is that I.V. was malingering. Malingered stuttering is probably not very common. Cases of malingered stuttering have been reported by Shirkey (1987) and Bloodstein (1988). One will remember that I.V. had complained about not receiving more individual therapy. It is not unthinkable then that she should have started to stutter in order to obtain again the individual attention she was receiving before she was enrolled for group therapy. However, the nature of her dysfluency was very consistent with patterns of genuine stuttering. She stuttered on consonants, not on vowels; as a rule she stuttered on syllable-intial consonants, and in polysyllable words it was most often the intial syllable of the word that was affected. On the other hand she did not produce repetitions of any kind. Yet, repetition is the symptom of stuttering that is probably best known to non-professionals and as such the symptom that one would certainly expect in malingered stuttering. It is very unlikely, therefore, that I.V. should have been malingering.

CONCLUSION

As the present case illustrates, several hypotheses should be considered in the diagnosis of a dysfluency that developped beyond the typical childhood period. The dysfluency may be a recurrence of childhood stuttering (see this volume, pp. 171-184), it may be drug-induced, it may be of neurogenic origin, it may be of psychogenic origin and one should even consider the possibility of malingered stuttering. Even when the stuttering occurred following brain damage, one should still consider possibilities other than neurogenic stuttering. As in the case of I.V. the stuttering could have been interpreted as neurogenic. There were even stronger suggestions, however, that her stuttering was psychogenic.

REFERENCES

ANDREWS G., QUINN P.T., SORBY W.A. (1972) Stuttering: an investigation into cerebral dominance for speech. *Journal of Neurology, Neurosurgery, and Psychiatry* 35: 414-418.

ARDILA A., LOPEZ M.V. (1986) Severe stuttering associated with right hemisphere lesion. *Brain and language* 27: 239-246.

BLOODSTEIN O. (1988) Verification of stuttering in a suspected malingerer. *Journal of Fluency Disorders* 13: 83-88.

BLOODSTEIN O. (1995) *A handbook on stuttering*. London: Chapman & Hall.

DEAL J.L. (1982) Sudden onset of stuttering: a case report. *Journal of Speech and Hearing Disorders* 47: 301-304.

DEAL J.L., DORO J.M. (1987) Episodic hysterical stuttering. *Journal of Speech and Hearing Disorders* 52: 299-300.

DEELMAN B.G., KONING-HAANSTRA M., LIEBRAND W.B.G., VAN DE BURG W. (1981) *S.A.N. Test, een afasietest voor auditief taalbegrip en mondeling taalgebruik.* Lisse: Swets & Zeitlinger.

ELLIOTT R.L., THOMAS B.J. (1985) A case report of alprazolam-induced stuttering. *Journal of Clinical Psychopharmacology*, 5: 159-160.

FLEET W.S., HEILMAN K.M. (1985) Acquired stuttering from a right hemisphere lesion in a right-hander. *Neurology* 35: 1343-1346.

GRAETZ P., DE BLESER R., WILLMES K. (1992) *De Akense afasietest. Nederlandstalige versie*. Lisse: Swets & Zeitlinger.

HELM-ESTABROOKS N. (1986) Diagnosis and management of stuttering in adults. In O ST. LOUIS K. *The atypical stutterer*. Principles and practices of rehabilitation. Orlando: Academic Press, 193-217.

HELM-ESTABROOKS N. (1993) Stuttering associated with acquired neurological disorders. In CURLEE R.F. (Ed.) *Stuttering and related disorders of fluency*. New York: Thieme Medical Publishers, 205-219.

HELM N.A., BUTLER R.B., BENSON F. (1978) Acquired stuttering. *Neurology* 28: 1159-1165.

HELM N.A. , BUTLER R.B., CANTER G.J. (1980) Neurogenic acquired stuttering. *Journal of Fluency Disorders* 5: 55-68.

INTERNATIONAL PHONETIC ASSOCIATION (1974) *The principles of the international phonetic association*. London: IPA.

LEBRUN Y. , BIJLEVELD H., ROUSSEAU J.J. (1990) A case of persistent neurogenic stuttering following a missile wound. *Journal of Fluency Disorders* 15: 251-258.

MAHR G., LEITH W. (1992) Psychogenic stuttering of adult onset. *Journal of Speech and Hearing Research* 35: 283-286.

MARKET K.E., MONTAGUE J.C., BUFFALO M.D., DRUMMOND S.S. (1990) Acquired stuttering. Descriptive data and treatment outcome. *Journal of Fluency Disorders* 15: 21-33.

McCLEAN M.D., McLEAN A. (1985) Case report of stuttering acquired in association with phenytoin use for post-head-injury seizures. *Journal of Fluency Disorders* 10: 241-255.

NURNBERG H.G., GREENWALD B. (1981) Stuttering: an unusual side effect of phenothiazines. *American Journal of Psychiatry* 138: 386-387.

QUADER S.E. (1977) Dysarthria: an unusual side effect of tricyclic antidepressants. *British Medical Journal*, 9: 97.

ROSENBEK J.C. (1984) Stuttering secondary to nervous damage. In CURLEE R., PERKINS W. (Eds.) *Nature and treatment of stuttering: new directions*. San Diego: College-Hill, 31-48.

ROTH C.R., ARONSON A.E., DAVIS L.J. (1989) Clinical studies in psychogenic stuttering of adult onset. *Journal of Speech and Hearing Disorders* 54: 634-646.

SHIRKEY E.A. (1987) Forensic verification of stuttering. *Journal of Fluency Disorders* 12: 197-203.

SILVERMAN F.H. (1992) *Stuttering and other fluency disorders*. Englewood Cliffs, New Jersey: Prentice-Hall.

STEWART T., GRANTHAM C. (1993) A case of acquired stammering: the pattern of recovery. *European Journal of Disorders of Communication* 28: 395-403.

TIPPETT D.C., SIEBENS A.A. (1991) Distinguishing psychogenic from neurogenic dysfluency when neurologic and psychologic factors coexist. *Journal of Fluency Disorders* 16: 3-12.

VAN DONGEN H.R., VAN HARSKAMP F., LUTEIJN F. (1976) *Tokentest. Handleiding.* Nijmegen: Berkhout.

VAN RIPER C. (1971) *The nature of stuttering*. Englewood Cliffs, New Jersey: Prentice-Hall.

H. BIJLEVELD, A.-M. SIMON

A CASE OF ACQUIRED STUTTERING FOLLOWING BRAIN DAMAGE

INTRODUCTION

Whilst neurogenic stuttering is by now a well-recognized nosological entity, its pathophysiology is still a matter of conjecture (see Lebrun, this volume). A case of acquired stuttering following brain damage sustained in adulthood is reported which may shed some light on the pathogenesis of the condition.

THE PATIENT

The patient is a right-handed French-speaking lawyer. In 1988 at the age of 44 he had a car accident which involved thoracic lesions, skull fracture and otorrhagia. He was comatose for 8 days and had to be ventilated for 9 days. Some problems with ventilation seem to have caused temporary brain hypoxia. NMR and CT scan were not contributive, though. Consciousness was gradually recovered. The patient started to speak but evidenced severe stuttering. He had never stuttered before. There was no obvious language disorder and neurological examination was normal. A few weeks later, the patient was discharged home. Ambulatory speech therapy was started to remedy the loss of fluency.

Some five months later, the patient was re-examined. He complained of thoracic pain, headaches, fatiguability, concentration problems, intolerance to noise and light, difficulty in remembering recent events and in performing calculations, slowness in writing and inability to follow group conversations and to play the piano. Neurological, electroencephalographic and neuroradiologic examinations failed to disclose anomalies. However, recordings of visual, auditory and somesthetic evoked potentials revealed a conduction disturbance in the white matter of the left hemisphere. The patient stuttered in spontaneous, provoked, imitative and serial speech as well as in oral reading and singing. There was no adaptation effect. The stuttering symptoms consisted primarily of repetitions.

On the other hand, there was no aphasia, apart from a few word-finding difficulties in the naming-on-confrontation task.

Verbal humour was preserved. However, short term verbal memory was reduced. The patient's score on Raven's matrices indicated an IQ of about

Y. Lebrun (ed.), From the Brain to the Mouth, 163–169.

120. Graphomotricity was slow. Reaction time to auditory and visual stimuli was increased. The patient could not clap his hands in accordance with a given rhythmic pattern nor produce the sound /a/ in accordance with the the beats of a metronome. He could no longer play the piano because of a lack of bimanual coordination. The performance of quick alternate hand movements was also disturbed.

A month later, the patient was videotaped during a variety of verbal tests. Analysis of the recordings showed that the patient's stutter consisted of sound, part-word, word, and word group repetitions. Sound repetitions occurred primarily in word initial position and affected consonants as well as vowels. Most part-word repetitions occurred also in word initial position. Not a single final sound or part-word repetition was noted. Word repetitions affected nearly always function words, such as *que*, *quand*, *très*, *pour* or *je*. Most of the time repeated word groups consisted of function words, e.g. *je me*, *ou des*, *que je*. Repetitions tended to be clonic when they comprised 3 or more reiterations of the speech segment.

In addition to involuntary repetitions there were occasional sound (particularly vowel) prolongations and also a number of blocks. The filler *euh* occurred fairly frequently in conversational speech. Besides, there were unexpected long pauses not only in conversational but also in serial or imitative speech. Indeed, serial speech was often slow, with pauses between the successive items of the overlearned series.

The following is an excerpt from the recorded conversation. In the transcription, a new line is begun whenever there was a pause in the patient's verbal output; dashes separate repeated speech segments; colons denote sound prolongations:

(Racontez-moi une journée normale ou une matinée normale)
Euh mn
euh alors
je m-j:e m-j:e m-euh-je m-je m-je me lève tôt
pour-pour-pour-pour deux raisons
euh ma-ma-ma fille Ma-Ma-
thilde se lève (laughs) et p-puis euh par-par-parfois pour-pour-pour venir ici
et euh puis par-parce-parce que depuis euh une-une-une dixaine de: jours je retourne un peu au-au-au bureau
bon alors
je m-je me lève
(sighs) bien-bien enfin la-la moitié du-d-d:u temps je-je prépare un-un pe-un-un-petit dé-dé-déjeuner

euh
m euh je-je-je-je dis quoi? (laughs).
(Par exemple?)
Je-je-je fais du:-du-du jus de fruit et frais ou des-ou-ou-des-des jus de:-de: lé-de légumes frais puis je-je f-f-fais griller des-des toasts (laughs) je-je prépare un-un café que-que-que je-que je ne bois pas euh et puis euh nous-nous
déjeunons.

Stuttering was present in all forms of verbal output including whispered oral reading and singing. When singing the patient made part-word and whole word repetitions. In addition he sang out of tune. On moments of stuttering, he tended to close his eyes. While stuttering obtained in all speaking situations, its severity was to some extent influenced by the degree of propositionality of the verbal output: dysfluencies were more frequent in conversational than in repetitive, completive or serial speech. Reading aloud was nearly as disturbed as conversational speech, though. Whispered oral reading was more dysfluent than voiced oral reading: 50% vs. 33% dysfluencies. Table 1 shows the relative frequency of the dysfluencies in various verbal tasks.

The patient was aware of his speech impediment and worried much about it. Yet, in conversation and during verbal testing he at times laughed unexpectedly. It is not clear whether these laughs were unintentional or were meant to cover up speaking difficulties or to lessen the tension caused by speaking efforts.

The patient was interviewed again 7 years after the onset of stuttering. The interview was tape-recorded and the patient's conversational speech analysed. Repetitions were more frequent than blocks and prolongations. Sound and part-word repetitions occurred primarily in word initial position, occasionally in the middle of a word but never at the end of words. In conversational speech, oral reading, repetition of sentences and the completion of idioms tasks, repetitions occurred more frequently on function than on content words. Repetitions comprised rarely more than two reiterations of the speech segment. Prolongations occurred at times within repetitions. The most frequent fillers were *euh*, *bon*, *enfin* and *quoi*. Moreover, an embololalic *voilà* occurred at the end of a number of utterances.

Table 2 shows the relative frequency of the various types of dysfluencies in conversational speech.

The patient maintained eye contact throughout the conversation. He stated that he stuttered as much during solitary monologues and solitary oral reading as during conversation or oral reading in front of an audience.

TABLE 1

	Conversation	Counting	Oral Reading	Repetitions		Completion of idioms	Total
				words	sentences		
Sound repetitions	4	1	2	0	0	1	8
Part-word repetitions	6	2	9	4	2	1	24
Word repetitions	15	4	24	0	5	12	60
Wordgroup repetitions	5	0	8	0	1	1	15
Total repetitions	30	7	43	4	8	15	107
Prolongations	4	0	6	0	2	1	13
Blocks	2	3	3	14	0	0	22
Fillers	12	0	3	0	0	0	15
Total syllables	118	53	154	50	37	65	477
Total dysfluencies	48	10	55	18	10	16	157
Percentage dysfluencies	41	19	36	36	27	25	33

TABLE 2

	Conversation
Sound repetitions	2
Part-word repetitions	5
Word repetitions	27
Wordgroup repetitions	8
Total repetitions	42
Prolongations	9
Blocks	6
Fillers	15
Total syllables	472
Total dysfluencies	72
Percentage dysfluencies	15

However, his stutter could fluctuate independently of his verbal activity. At certain times he spoke more, and at other times less, fluently. The patient subjectively felt that, on the whole, he stuttered less than during the first two or three years post onset. Comparison of Table 1 and 2 supports this impression. The patient ascribed the improvement to the many oral reading exercises which he had done and also to his re-acquisition of piano playing skills. Indeed, he was convinced that re-learning to play the piano had had a favourable influence on his speech.

DISCUSSION

This patient, then, had never been dysfluent prior to the brain trauma he suffered at age 44. His stutter can therefore be regarded as an acquired speech disorder. As it was present as soon as the patient recovered consciousness, it can hardly have been psychological in nature. Indeed, if it had come about as a psychic reaction to the post-traumatic deficits, it would have taken some time to develop. It would have emerged after the patient had returned home and had come to realize that the accident had caused some cognitive deficits. In this case then, the stutter appears to have resulted directly from the cerebral trauma, i.e. to have been neurogenic in nature. This conclusion is important in view of the fact that a few months after the accident, neurological, electroencephalographic and neuroradiologic examinations were normal. This shows that brain damage can cause a number of cognitive-attentional, motor skills and speech deficits and yet be undetectable by means of clinical and paraclinical investigations.

When examined six months post-onset, the patient stuttered in all proposed verbal tasks including whispered reading. The omnipresence of the dysfluency is important in view of Deal's suggestion (1982) that a stutter which persists during whispered speech is likely to be psychogenic. The present case shows that neurogenic stuttering may very well affect all forms of oral-verbal output including whispered speech.

It may be of some significance that the brain trauma impaired both the ability to speak fluently and the ability to play the piano. In the course of time, either ability improved. Indeed, the patient subjectively felt that the exercises he did to recover his musical agility benefitted his speech delivery.

The patient never lost his ability to read or appreciate music. He did not lose his linguistic competence either. As a matter of fact, he at no time evidenced aphasia. But his musical and linguistic motor skills were affected: they had lost their fluidity. The mechanics of speech production and piano playing were disturbed. As argued by Lebrun in this volume (pp. 129-130) there are reasons to implicate the brainstem in the occurrence of acquired stuttering. The present case suggests that piano playing as a much-practised motor skill may also be largely dependent on sub-cortical structures. Speech fluency and musical fluency may both be based on routines that develop within the extrapyramidal system. Damage suffered by this system is therefore likely to impair either ability, as in the present case. And it is not unthinkable that retraining one may have a beneficial influence on the recovery of the other, as the patient maintained.

The present case may be compared with the one reported by Cipolotti et al. (1988). Their patient had acquired stuttering following damage to the basal ganglia. He had no aphasia, no apraxia and no cognitive impairment. But he could no longer perform prescribed sequences of key depressing finger movements.

Helm et al. (1980) have stated that neurogenic acquired stuttering is often concomitant with such non-verbal neuropsychological deficits as difficulty reproducing and sustaining alternating sequential hand positions, difficulty singing melodies, and difficulty reproducing rhythmic tapping patterns. These three deficits were also observed in the present case. Helm et al. further expressed the view that difficulty drawing three dimensional figures and copying block designs is a frequent concomitant of neurogenic stuttering. This deficit was not noted in the present case, however.

It appears that while adult-onset stuttering following brain damage may be accompanied by a number of neuropsychological problems, these problems may vary from patient to patient.

CONCLUSION

The present case then shows that neurogenic stuttering may obtain in the absence of neuroradiologically demonstrable brain damage. Moreover, the speech impediment may be concomitant with a disorder of purposeful production of sequences of quick finger movements. Both impairments may possibly be due to extrapyramidal dysfunctioning.

REFERENCES

CIPOLOTTI L., BISIACCHI P., DENES G., GALLO A. (1988) Acquired stuttering: A motor programming disorder? *European Neurology* 28: 321-325.

DEAL J.(1982) Sudden onset of stuttering: A case report. *Journal of Speech and Hearing Disorders* 47: 301-304.

HELM N., BUTLER R., CANTER G. (1980) Neurogenic acquired stuttering. *Journal of Fluency Disorders* 5: 269-279.

J. VAN BORSEL, V. CAPPAERT

AGGRAVATION OR RECURRENCE OF PRE-EXISTING STUTTERING FOLLOWING BRAIN DAMAGE SUFFERED IN ADULTHOOD

INTRODUCTION

Stuttering is a speech disorder which typically begins in childhood, usually between the ages of 2 and 5. Occasionally, however, stuttering is also observed for the first time in adulthood. In the latter case the stutter appears often secondary to acquired brain damage. Sometimes it is also a side effect of medication or it may be of psychogenic origin related to some form of psychological stress. In order to distinguish between the two conditions, stuttering which has its onset in childhood is often called *developmental stuttering* while stuttering originating after the typical childhood period is called *acquired stuttering.*

Sometimes a more complex situation arises. This is the case when a patient with developmental stuttering suffers brain damage in adulthood and as a consequence of the damage shows recurrence or aggravation of the pre-existing dysfluency. As there is in such cases a history of developmental stuttering, the dysfluency appearing after the neurologic episode cannot really be called acquired stuttering. On the other hand it cannot be classified as developmental stuttering either. If the patient had not suffered brain damage his speech would not have looked the way it does.

Recurrence or aggravation of developmental stuttering after brain damage suffered in adulthood is not a frequent condition and only a small number of such cases have been described in the literature. These cases are interesting, though, as they may provide clues to the mechanism underlying developmental stuttering. This paper reports the aggravation of pre-existing dysfluency in an adult after bilateral brain damage. Findings are compared with the other existing reports on aggravation or recurrence of developmental stuttering and are discussed in the light of hypotheses on the etiology of stuttering.

CASE HISTORY

On May 18, 1993, J.B., a 49-year-old right-handed male was rushed to the Ghent university hospital. He had collapsed while gardening. When his wife found him he was unconscious. Admittance CT and a cerebral

Y. Lebrun (ed.), From the Brain to the Mouth, 171–184.

arteriogram revealed that J.B. had suffered a subarachnoid hemorrhage due to rupture of an aneurysm of the right arteria communicans posterior. Both cerebral hemispheres were involved and the bleeding extended from the frontal to the parietal regions. Because of the risk of angiospasm and cerebral oedema clipping of the aneurysm was initially delayed. A few days later, however, (June 13, 1993) compressive problems (compression of the nervus opticus) indicated the need for urgent operative treatment to prevent rebleeding. As is often the case in subarachnoid hemorrhage hydrocephaly developed. Because the raised intracranial pressure could not be reduced by a ventricular catheter a permanent ventriculo-atrial shunt was implanted on June 29, 1993.

In August 1993 the patient was admitted to the hospital rehabilitation center. The clinical neurological examination on admission disclosed post-traumatic amnesia, disorientation in space, time, person and condition, and memory problems. Ocular movements and pupillary reactions were normal but there was facial paresis on the left. Exteroceptive sensation was normal, proprioception was impaired in both hands and feet. There were no symptoms of cerebellar deficiency. Muscle tone was normal and reflexes too were normal and symmetrical. No Babinski sign nor any clonus were present. Muscular power in the upper as well as the lower extremities was 4/5. There were no muscular contractures.

As far as the patient's speech was concerned no stuttering was observed at that time. The patient was dysphonic, however, and remained so for several weeks. Also, he complained of a decreased hearing acuity on the right.

An ENT examination in November 1993 revealed normal laryngeal structures with incomplete closure of the true vocal folds and a pronounced ventricular dysphonia. Audiometric assessment disclosed a sensorineural hearing loss (−60 to −80 dB) on the right presumably as the result of some cochlear trauma (A BERA ruled out the possibility of a retrocochlear pathology and on NMR there was no evidence of an acoustic neuroma).

At the end of September 1993 the patient was administered a number of subtests of an experimental Dutch version of the Boston Diagnostic Aphasia Examination (Goodglass and Kaplan, 1972). Results of this testing (see Table 1) indicated the presence of a mild aphasia. Auditory comprehension was fairly good but not normal though. Especially the comprehension of complex ideational material proved difficult. Oral expression was nearly normal but there was a reduced fluency in controlled association as tapped by the subtest "animal naming". Understanding of written language was

TABLE 1
Boston diagnostic aphasia examination

AUDITORY COMPREHENSION	
Word discrimination:	60/72 (83%)
Body part identification:	18/20 (90%)
Commands:	13/15 (86%)
Complex ideational material:	5/12 (41%)
ORAL EXPRESSION	
Oral agility: nonverbal agility:	11/12 (91%)
: verbal agility:	14/14 (100%)
Automized sequences:	6/6 (100%)
Repetition of words:	10/10 (100%)
Repeating phrases: high probability:	8/8 (100%)
: low probability:	8/8 (100%)
Word reading:	27/30 (90%)
Responsive naming:	30/30 (100%)
Visual confrontation naming:	96/96 (100%)
Body-part naming:	24/30 (80%)
Animal-naming:	8*
Oral sentence-reading:	10/10 (100%)
UNDERSTANDING WRITTEN LANGUAGE	
Symbol and word discrimination:	9/10 (90%)
Phonetic association: word recognition:	8/8 (100%)
: comprehension of oral spelling:	6/8 (75%)
Word-picture matching:	9/10 (90%)
Reading sentences and paragraphs:	9/10 (90%)
WRITING	
Mechanics of writing:	3 (normal)
Recall of written symbols: serial writing:	47/47 (100%)
: primer level dictation:	15/15 (100%)
Written word finding: spelling to dictation:	10/10 (100%)
: written confrontation naming:	9/10 (90%)
Written formulation: narrative writing:	**
: sentences written to dictation:	4 (normally written)

* there is no maximum score for this subtest
** not tested

also fairly good except for the comprehension of oral spelling. Writing was almost completely normal.

Neuropsychological testing in December 1993 (WAIS) revealed a well-preserved verbal IQ (VIQ 96) but a reduced performance IQ (PIQ 66).

In the rehabilitation center J.B. received intensive physical therapy, occupational therapy and speech language therapy as well as neuropsychological guidance. The speech language therapy concentrated on the voice problem, the deficits in orientation and memory and the comprehension difficulties.

We saw J.B. for the first time on March 29, 1994. It was noted that since a few weeks the patient had presented stuttering symptoms. The dysfluency had gradually worsened and the consulting physician of the rehabilitation center wondered how this stuttering behaviour fitted in the overall clinical picture.

An initial session confirmed that the patient demonstrated stuttering behaviour. His speech was dysfluent and was characterized by numerous prolongations, blocks and repetitions. The prolongations and blocks were sometimes of considerable length and not infrequently even contained a respiratory pause. When after some trials the patient did not succeed in uttering a word he would sometimes start to write the word with the finger in the air. Sometimes he also named the letters of the intended word.[1] J.B was very worried about his stuttering and on several occasions during the intitial session he burst into tears. He was particulary anxious that people should start laughing at his stutter.

Asked for the onset of his dysfluency the patient reported that he had always stuttered since childhood. He was not very precise, however, on this point. At one moment he recalled scenes of mockery from the time he was still in primary school. Shortly thereafter he reported that his stuttering had begun all of a sudden in secondary school when one day he had been accused unjustly of unchastity by a teacher-priest.

Immediately after the initial session we had an interview with the patient's spouse. Worrying about her husband's dysfluency she had already asked his relatives about the existence of any childhood stuttering. She was told that her husband had stuttered as a toddler but that he had outgrown the problem during primary school. Certainly he was never given any speech therapy. She herself denied ever having remarked genuine stuttering in her husband until recently. She had observed, though, as soon as they got acquainted, that from time to time J.B. slightly paused, usually at the beginning of a word and most often on words with an intial /p/. It occurred to her that she had noticed this behaviour most often when her husband spoke on the telephone. As far as J.B.'s present dysfluency was concerned, she was the one who had suggested him to recur to writing or spelling aloud when he did not succeed in producing a word otherwise.

[1]He succeeded in naming the letters of words which he could not utter because the alphabetic name of a letter is often different from the sound that is represented by the letter.

Before he suffered his stroke J.B. was employed as an educator in an institute for the deaf. During the initial session the patient mentioned that his stuttering did not really hinder him in his professional activities. He sometimes stuttered when he had to speak on meetings but he was never laughed at. Yet, in a later session J.B. reported that he had received therapy for his stuttering in the institute where he was employed. As J.B.'s wife appeared uninformed of any therapy we contacted a colleague at the institute who had known the patient for several years. He formally denied that J.B. had been given speech therapy. He knew that J.B had always been somewhat dysfluent but not to a high degree. He described J.B.'s dysfluencies as "very mild".

On May 31, 1994 and June 7, 1994 video-recordings were made of J.B's speech and subjected to analysis. Examination of conversational speech showed that J.B.'s dysfluency was predominantly characterized by prolongations and blocks. In a randomly selected 620 word sample with a total dysfluency index of 12.1%, prolongations and blocks accounted for respectively 49.3% and 18.7% of the total number of dysfluencies. The prolongations and blocks had duration times that varied highly, however, from one stuttering moment to another. At times a prolongation or block hardly lasted a second, at other times its duration exceeded over half a minute. Both prolongations and blocks usually occurred on consonants. Blocks occurred exclusively on plosives, prolongations most often on fricative sounds and a number of times also on a liquid, a glide or a nasal. On closer inspection some of the instances of prolongation of a nasal actually appeared to be attempts to produce a target plosive. Attempting a word with an initial /d/ the patient would sometimes produce a prolongation of the homorganic nasal /n/, or instead of a target /b/ he would prolong an /m/. Vowels and diphthongs were only rarely stuttered on and the few instances of dysfluency that did occur with these types of sounds were slight prolongations.

In addition to prolongations and blocks J.B.'s conversational speech also contained instances of repetition of phrases, words and part-words. These types of dysfluency, which more often than not occurred within strings of very hurried, tachylalic speech, were not as frequent, however. In the 620 word sample that was further analysed repetitions of part-words, words and phrases accounted together for only 17.3% of the total number of dysfluencies. Moreover, compared to certain prolongations and blocks the dysfluencies resulting from repetition were always very mild. At most a sound, syllable, word or phrase was repeated two or three times.

The remainder of J.B.'s dysfluencies consisted of revisions, incomplete phrases and interjections. These categories accounted for respectively 5.3%, 6.7% and 2.7% of the total number of dysfluencies.

A further analysis of the loci of the prolongations, blocks and part-word repetitions revealed that these dysfluencies always occurred in syllable initial position. Their occurrence was not restricted, however, to the *first* syllables of words and instances of prolongations, blocks and repetitions in word medial position were found as well. Neither was the occurrence of prolongations, blocks and part-word repetitions restricted to words of high information value. On the whole function words appeared no less vulnerable than substantive words.

In addition to J.B.'s conversational speech we also studied his speech during a repetition task, a sentence completion task, a reading task and while singing.

For testing fluency during repetition the patient was administered the subtest "Naspreken" from the Dutch version of the Aachener Aphasie Test (Graetz, De Bleser and Willmes, 1992). This test comprises three parts: sound repetition, word repetition and sentence repetition. No dysfluencies were observed during sound repetition. Also speech during repetition of sentences (totalling 71 words) was almost completely fluent. Apart from one word repetition and one prolongation of an initial /h/ no stuttering occurred. During the repetition of single words, however, the dysfluency index amounted to 60.7%. Even more than in conversational speech prolongations and blocks were the predominant dysfluency type and constituted 16 out of the total of 17 stutters observed. Once a part-word repetition was noted.

Stuttering was also observed during the sentence completion test but not as frequently. This test, which was a Flemish adaptation of the original test from the Cooper Personalized Fluency Control Therapy (Cooper and Cooper, 1990), required the patient to complete 12 sentences with in all 18 words. 4 words (i.e. 22%) were stuttered on. The dysfluencies were 2 prolongations, 1 block and 1 part-word repetition.

As material for the reading task the Dutch version of the IPA-text "The north wind and the sun were disputing which was the stronger . . .". (International Phonetic Association, 1974) was used. We had the patient read this text (comprising 120 words in Dutch) five times in succession. The amount of stuttering was limited and no adaptation effect could be discerned, the number of stuttered words for the five readings being respectively 5, 3, 5, 5 and 3. There was some consistency in J.B.'s dysfluency during reading in that two words were affected on three of the five readings and another two on two of the five readings. Still, eleven other words were all involved only once. Again the majority of dysfluencies were prolongations and blocks.

When asked for his leisure interests during the interview we had for the purpose of collecting a sample of conversational speech, J.B. told us that he was fond of singing Gregorian chants. He was very willing then to demonstrate some chanting. On request he also sang part of a popular Dutch song. No dysfluencies were noted during singing.

Finally we also determined the severity of J.B.'s stuttering using the procedure outlined by Riley (1972) and assessed his attitudes toward interpersonal communication by means of the Erickson S-scale (Erickson 1969). Evaluation according to the Stuttering Severity Instrument of Riley classified J.B. as a very severe stutterer with a total overall score of 41 (frequency task score 15, duration score 18, total physical concomitant score 8). Assessment with the Erickson S-scale showed that notwithstanding his severe stuttering, overall J.B. had a positive attitude towards speaking, scoring on Percentile 53 compared to healthy speakers.

J.B. remained at the Ghent university hospital rehabilitation center for a number of weeks until he was transferred to another center. He still stuttered severely then. About six months later we had contact with his wife again. She told us that her husband's speech had somewhat improved but that the stutter was still present.

DISCUSSION

Van Riper (1971) pointed out that in cases of adult onset stuttering "there is always some suspicion or even evidence to indicate that the adult stuttering was a recurrence of a problem which had shown itself much earlier in the person's life" (p. 62). He further warned that one must evaluate accounts of adult-onset stuttering with some caution and that "when carefully investigated, some of them (the patients) turn out to be interiorized stutterers who under great stress, could no longer hide their disorder" (pp. 69, 70). In accordance with Van Riper's warning it has been a concern of most investigators reporting cases of adult-onset stuttering to exclude any possible doubt about a previous history of developmental stuttering. The literature on adult onset stuttering is almost exclusively devoted then to the study of what one could call "pure" cases of acquired stuttering. The possible influence of brain damage on the speech dysfluency of developmental stutterers has not been studied systematically.

As the present case illustrates, brain damage may at any rate significantly alter a stutterer's dysfluency. Our patient showed a pre-existing mild dysfluency which he had probably learned to control to some extent. Neurological damage caused a marked aggravation of this dysfluency. Similar cases of aggravation of previous stuttering have been reported by Mazzucchi et al. (1981) and Riggs et al. (1983). In addition, it has also been reported that following a brain lesion stuttering may reappear in those patients who only transiently stuttered during childhood or who were completely cured of a childhood dysfluency (Rosenbek et al. 1978, Mazzucchi et al. 1981, Helm-Estabrooks et al. 1986).

The type of neurological insult responsible for the aggravation of the dysfluency in our patient was a cerebrovascular accident. This is in line with most of the other cases reported thus far in which it was also a cerebrovascular accident that caused the recurrence or aggravation of a pre-existing dysfluency. Exceptions exist, however. One of the patients studied by Mazzucchi et al. (1981), case 5, a 25-year-old male, showed aggravation of developmental stuttering as a sequel of trauma with coma in the absence of any detectable lesion. And a patient of Quinn and Andrews (1977), a 62-year-old businessman who had stuttered transiently during childhood, presented again with stuttering secondary to a dementing process, probably Alzheimer's disease. There is some doubt, however, whether in the latter case it was really the previously existing stuttering that had re-emerged. The authors concluded somewhat equivocally that in this case "stuttering similar in form to his earlier childhood stutter, was presumably the reiterative utterance of parts of words which is so characteristic of Alzheimer's disease" (p. 700).

In the present case dysfluency aggravated following a bilateral brain lesion. Involvement of both cerebral hemispheres does not seem to be a conditio sine qua non, however, for a pre-existing dysfluency to re-emerge or to aggravate. In fact, in most of the cases reported to date there was a unilateral lesion. As a rule the lesion appeared to be situated in the left hemisphere but not, however, in any particular area. In Rosenbek et al.'s (1978) patient, for instance, the lesion was localized in the frontal and temporal lobe whereas in Riggs et al.'s (1983) patient a parietal lesion was found.

While there was in all cases reported thus far either a left-sided or bilateral lesion and while apart from the case of Helm-Estabrooks et al. (1986) patients were all righthanded, stuttering following brain damage was not always concomitant with aphasia. In some cases there was definitely aphasia (Rosenbek et al. 1978; Mazzucchi et al. 1981, case 10). In others, however, there were only transient aphasia signs in the acute stage (Mazzucchi et al. 1981, case 7; Riggs et al. 1983), and in still others no aphasia was found at all (Mazzucchi et al. 1981, case 5 and 14). Interestingly, also Helm-Estabrooks et al.'s (1986) left-handed patient belonged to the last group. Before his left-sided stroke this patient wrote with his right hand because he was forced to at school, but he used his left hand for all other activities. There was moreover a familial tendency towards left-handedness. In this patient, then, the right hemisphere was perhaps dominant for speech.

Cases of re-emergence or aggravation of developmental stuttering further seem to differ as to the onset and duration of the dysfluency following neurological damage. In the present case the pre-existing dysfluency started to develop gradually some nine months post-onset. Six months later the patient still stuttered. A gradual development of dysfluency was also seen in case 5 of Mazzucchi et al. (1981). Yet, a sudden onset concurrently with the neurological episode appears to be the more common pattern. On the other hand, the persistence of the dysfluency observed in our patient conforms to the usual picture. In some cases dysfluency was seen to persist for 3 to 5 years after the triggering neurological insult (Rosenbek et al. 1978; Mazzucchi et al. 1981, case 14). But again, a different course seems possible. The patient of Riggs et al. (1983) showed frequent repetitions, hesitations and also sound prolongations at onset but within 24 hours after admission his stutter had already noticeably improved. Forty-eight hours after admission he evidenced only a mild degree of stuttering and two weeks later no major dysfluencies were found.

Finally, the present case confirms the trend evident from the available reports that re-appearence or aggravation of a previously existing dysfluency is most likely to be observed in male victims. Actually, we know of no similar observations in females. This is of course not totally surprising, knowing that developmental stuttering is three to four times more frequent in males than in females.

Remarkably enough brain damage may apparently change a developmental stutterer's dysfluency in two directions. Quite contrary to the present case and the cases of recurrence or aggravation of pre-existing stuttering discussed above, some developmental stutterers became, at least temporarily, more fluent after a brain lesion. This was for instance the case in the first patient described by Helm-Estabrooks et al. (1986). This patient, a man who began to stutter at age 8, no longer stuttered when he regained consciousness 10 days after a head injury. Reexamination 1 year post-onset showed that he then again produced more dysfluencies but that his speech was still more fluent than it had been before the accident. Remission of stuttering was also observed by Cooper (1983) in a 25-year-old girl who was a chronic severe stutterer. Following a brain-stem contusion the girl experienced a 6-week period of normal fluency. Subsequently she remained relatively fluent for 6 months until her stuttering began to return to its former state. Also Miller (1985) reported cessation of dysfluency in two developmental stutterers. His patients were two men developing multiple sclerosis in whom stuttering abated as other forms of speech impairment appeared.

Improvement of speech has further been observed in a number of patients who underwent neurosurgery. Four such patients were described by Jones (1966). All four were lifelong stutterers who developed focal brain lesions requiring surgery. In one patient a tumour was excised, in the three others an aneurysm was clipped. Postoperatively none of these patients continued to stutter. Follow-up ranged from 15 months to 3 years. Similar cases of cessation of stuttering after neurosurgery have also been reported by Guillaume et al. (1957) and Mazars et al.(1970). In their patients stuttering ceased completely after a total or partial temporal lobectomy.

Clearly, then, brain damage may either cause a developmental dysfluency to reappear or aggravate or it may on the contrary alleviate or abolish the stutter. How are these phenomena to be explained?

One hypothesis is that, at least in the case of recurrence or aggravation of a previous dysfluency, the change is emotionally based. It is well-known that stutterers become more dysfluent under great stress. Could it not be then that a dysfluency re-emerges or aggravates under the stress conceivably brought about by the neurological insult? Perhaps this hypothesis may hold for some cases. For others, however, it certainly does not. In the case reported by Riggs et al. (1983), for instance, the patient had experienced acute onset of stuttering in absence of any other neurological symptoms. As the re-appearence of a childhood stutter, which the patient had learned to control, was the only symptom he complained of, the dysfluency can hardly be explained as an emotional reaction. Riggs et al. (1983) are of the opinion that the cerebral insult (CT revealed a small, left-parietal hematoma) interfered with cerebral pathways that allowed the patient to control the stutter, which, they think, might represent a genetically inherited neurologic disorder. Stuttering, they believe, is caused by a genetic abnormality of the central nervous speech control systems.

The hypothesis of an emotionally based change is certainly problematic in cases of subsidence or cessation of stuttering after a brain lesion. Why should emotional stress have in some patients exactly the opposite effect to that so typically seen in stutterers? Two of the stutterers of Mazars et al. (1970) in whom neurosurgery was done under local anesthesia, started to speak fluently at the very moment of incision of some part of the brain.

Such an observation indeed militates against a psychological explanation for the cessation of stuttering. The explanation suggested by Lebrun and Bayle (1973) is that in these patients, who underwent a (partial) temporal lobectomy, auditory feedback was altered. It is well-known that stutterers tend to speak fluently when their auditory feedback is altered. Alteration of auditory feedback was also suspected in one of the cases of Jones (1966) in whom there was an occlusion of the left middle cerebral artery. Alternatively, for the other cases reported by Jones (1966) in whom there was a frontal lobe involvement, Lebrun and Bayle (1973) refer to evidence adduced by Eisenson (1958, pp. 232–235) of an abnormal inclination for perseverative behaviour in some stutterers. As it has been shown that lesions of the supplementary motor area and of the upper premotor region may bring about repetitive speech behaviour (Alajouanine et al. 1959; Arseni and Botez 1961), it is not unthinkable according to Lebrun and Bayle, that in these patients the neurosurgery indeed interfered with an abnormal propensity to perseveration, in the same way as stereotactic surgery interferes with unvoluntary movements. Lebrun and Bayle (1973) further suggested that the disruptive area may, but need not necessarily, be situated in the hemisphere that is not dominant for speech and that stuttering perhaps results from excessive interference of that area in the speech process.

Interhemispheric interference, be it in a different way, was also considered the possible cause for stuttering by Helm-Estabrooks et al. (1986). These investigators refer to the Orton–Travis theory (Orton, 1927; Travis, 1931) which they believe can explain the observation that in some developmental stutterers stuttering re-emerges after a brain lesion and that others experience remission of stuttering. According to Orton and Travis there is in stutterers a lack of cerebral dominance creating mistiming of the motor impulses to the bilateral speech muscles. Helm-Estabrooks et al. (1986) assume that in stutterers who have outgrown their childhood stuttering, dominance of one hemisphere has come about with maturation. Stuttering would recur in these patients when the dominance of that one hemisphere is again undone by a cerebral lesion. So, in recovered stutterers with left hemisphere dominance stuttering would recur after a left-sided lesion, in recovered stutterers with right hemisphere dominance stuttering would reappear after a right-sided lesion. A lesion in the non-dominant hemisphere, on the other hand, would only strenghten the unilateral dominance and stuttering would not re-occur. In stutterers who did not outgrow their childhood stuttering, there would still be interhemispheric competition with no clear dominant hemisphere. In these patients a unilateral lesion either in the left or in the right hemisphere would eliminate interhemispheric conflict and the stuttering would cease.

Lebrun has formulated yet another neurolinguistic theory on stuttering that could also explain the recurrence or aggravation and alleviation or abolishment of developmental stuttering (this volume, pp. 129-133). In this theory it is assumed that under normal circumstances the extrapyramidal system takes part in speech production under control of the cortex. In particular the more routinized aspects of speech such as articulation and rate of delivery would be taken care of by the extrapyramidal system. It is conjectured that developmental stutterers are individuals in whom the normal cortical control over the extrapyramidal participation in speech production has not been achieved. In neurogenic stuttering there would be either a dysfunction of the extrapyramidal system due to a subcortical lesion or there would be a hyperfunction of the extrapyramidal system due to a cortical lesion preventing the cortex from exercising adequate control over the extrapyramidal system. Similarly, a (further) reduction of cortical control over the extrapyramidal system might be presumed in cases of recurrence or aggravation of previously existing developmental stuttering. In cases of alleviation or abolishment of developmental stuttering following brain damage one could assume, at least in cases of trauma, that due to concussion or contusion the extrapyramidal system has become less able to resist cortical control. Alternatively, in cases such as the ones of Jones (1966) where a cerebral tumour was resected or an aneurysm was clipped one might assume that the surgical intervention enabled the cortex to function more normally again and to excersise improved control over the extrapyramidal system.

From the present case it is hard to tell which of the above theories is most likely to be the correct one. As in the case of Riggs et al. (1983) our patient seemed to have learned to control his stutter to some extent. It may be then that the hemorrhage he suffered disrupted central nervous pathways that allowed him to control his stutter. It is not impossible either that the damage which involved also the frontal lobes further extended an already present disruptive area interfering to some degree with the normal speech process. Neither is it unthinkable that the lesion brought on more interhemispheric competition. Though there was no unilateral but a bilateral lesion in our patient the effect of the lesion may nonetheless have been more outspoken in one hemisphere than in the other. On the other hand it is not impossible that the lesion brought about a further reduction of cortical control over the extrapyramidal system.

CONCLUSION

Following brain damage previously fluent speakers may sometimes present with stuttering. In developmental stutterers brain damage may alter speech in two opposite ways. In some patients a brain lesion alleviates or even abolishes the stutter. In others, like in our patient, the lesion causes a pre-existing dysfluency to recur or to aggravate. The mechanism behind these different patterns in stutterers is still unclear and warrants further investigation. It might be useful to study cases of recurrence or aggravation of dysfluency in stutterers following brain damage as a separate group, for the condition of these patients is different from both that of "ordinary" developmental stuttering and that of genuine acquired stuttering. Perhaps the dysfluency observed in this particular group of patients could be referred to as "renewed stuttering". Analogously one could speak of "renewed fluency" when stuttering completely ceased after a neurological episode. The latter condition too requires further investigation and may contribute to the search for the pathogenesis of stuttering.

REFERENCES

ALAJOUANINE T., CASTAIGNE P., SABOURAUD O., COTAMIN F. (1959) Palilalie paroxystique et vocalisations itératives au cours de crises épileptiques par lésion intéressant l'aire motrice supplementaire. *Revue Neurologique* 101: 685-697.

ARSENI C., BOTEZ M., (1961) Speech disturbances caused by tumours of the supplementary motor area. *Acta Psychiatrica et Neurologica Scandinavia* 36: 279-299.

COOPER E.B. (1983) A brain-stem contusion and fluency: Vicki's story. *Journal of Fluency Disorders* 8: 269-274.

COOPER E.B., COOPER C.S. (1990) *Cooper personalized fluency control therapy. Revised.* Arnhem, Boek- en leermiddelencentrale.

EISENSON J. (ed.) (1958) *Stuttering. A symposium.* New York, Harper and Row.

ERICKSON R.L. (1969) Assessing communication attitudes among stutterers. *Journal of Speech and Hearing Research* 12: 711-724.

GOODGLASS H., KAPLAN E. (1972) *Boston diagnostic aphasia examination (BDAE).* Philadelphia, Lea and Febiger.

GUILLAUME J., MAZARS G., MAZARS Y. (1957) Intermédiaire épileptique dans certains types de bégaiement. *Revue Neurologique* 96: 59-61.

GRAETZ P., DE BLESER R., WILLMES K. (1992) *De Akense afasietest. Nederlandstalige versie.* Lisse, Swets & Zeitlinger.

HELM-ESTABROOKS N., YEO R., GESCHWIND N., FREEDMAN M., WEINSTEIN C. (1986) Stuttering: disappearence and reappearence with acquired brain lesions. *Neurology* 36: 1109-1112.

INTERNATIONAL PHONETIC ASSOCIATION (1974) *The principles of the international phonetic association.* London, IPA.

JONES R. (1966) Observations on stammering after localized cerebral injury. *Journal of Neurology, Neurosurgery, and Psychiatry* 29: 192-195.

LEBRUN Y., BAYLE M. (1973) Surgery in the treatment of stuttering. In LEBRUN Y., HOOPS R. (eds.) *Neurolinguistic approaches to stuttering.* The Hague/Paris, Mouton: 82-89.

MAZARS G., HECAEN H., TZAVARAS A., MERIENNE L. (1970) Contribution à la chirurgie de certains bégaiements et à la compréhension de leur physiopathologie. *Revue Neurologique* 122: 213-220.

MAZZUCCHI A., MORETTI G., CARPEGGIANI P., PARMA M. (1981) Clinical observations on acquired stuttering. *British Journal of Disorders of Communication* 16: 19-30.

MILLER A.E. (1985) Cessation of stuttering with progressive multiple sclerosis. *Neurology* 35: 1341-1343.

ORTON S. (1927) Studies in stuttering. *Archives of Neurology and Psychiatry* 18: 671-672.

QUINN P.T., ANDREWS G. (1977) Neurological stuttering-a clinical entity? *Journal of Neurology, Neurosurgery, and Psychiatry* 40: 699-701.

RIGGS J.E., NELSON L.J., LANHAM J.E. (1983) Reemergence of stuttering after focal cerebral insult. *Archives of neurology* 40: 260.

RILEY G.D. (1972) A stuttering severity instrument for children and adults. *Journal of Speech and Hearing Disorders* 37: 314-322.

ROSENBEK J., MESSERT B., COLLINS M., WERTZ R.T. (1978) Stuttering following brain damage. *Brain and Language* 6: 82-96.

TRAVIS L.E. (1931) *Speech pathology.* New York, Appleton Century.

VAN RIPER C. (1971) *The nature of stuttering.* Englewood Cliffs N.J., Prentice-Hall.

INDEX

NEUROPSYCHOLOGY AND COGNITION

The purpose of the Neuropsychology and Cognition series is to bring out volumes that promote understanding in topics relating brain and behavior. It is intended for use by both clinicians and research scientists in the fields of neuropsychology, cognitive psychology, psycholinguistics, speech and hearing, as well as education. Examples of topics to be covered in the series would relate to memory, language acquisition and breakdown, reading, attention, developing and aging brain. By addressing the theoretical, empirical, and applied aspects of brain-behavior relationships, this series will try to present the information in the fields of neuropsychology and cognition in a coherent manner.

Series Editor: R. Malatesha Joshi, *Oklahoma State University, U.S.A.*

1. P.G. Aaron: *Dyslexia and Hyperlexia.* 1989
   ISBN 1-55608-079-4; 1994, Pb 0-7923-3155-9
2. R.M. Joshi (ed.): *Written Language Disorders.* 1991 ISBN 0-7923-0902-2
3. A. Caramazza: *Issues in Reading, Writing and Speaking.* A Neuropsychological Perspective. 1991 ISBN 0-7923-0996-0
4. B.F. Pennington (ed.): *Reading Disablities.* Genetic and Neurological Influences. 1991 ISBN 0-7923-1606-1
5. N.H. Hadley: *Elective Mutism.* A Handbook for Educators, Counsellors and Health Care Professionals. 1994 ISBN 0-7923-2418-8
6. W.C. Watt (ed.): *Writing Systems and Cognition.* Perspectives from Psychology, Physiology, Linguistics, and Semiotics. 1994 ISBN 0-7923-2592-3
7. I. Taylor and D.R. Olson (eds.), *Scripts and Literacy.* Reading and Learning to Read Alphabets, Syllabaries and Characters. 1994 ISBN 0-7923-2912-0
8. V.W. Berninger (ed.): *The Varieties of Orthographic Knowledge.* I: Theoretical and Developmental Issues. 1994 ISBN 0-7923-3080-3
9. C.K. Leong and R.M. Joshi (eds.): *Developmental and Acquired Dyslexia.* Neuropsychological and Neurolinguistic Perspectives. 1995
   ISBN 0-7923-3166-4
10. N. Gregg: *Written Expression Disorders.* 1995 ISBN 0-7923-3355-1
11. V.W. Berninger (ed.): *The Varieties of Orthographic Knowledge.* II: Relationships to Phonology, Reading, and Writing. 1995 ISBN 0-7923-3641-0
    Set (Vols 8 + 11) ISBN 0-7923-3081-1
12. Y. Lebrun (ed.): *From the Brain to the Mouth.* Acquired Dysarthria and Dysfluency in Adults. 1997 ISBN 0-7923-4427-8

KLUWER ACADEMIC PUBLISHERS – DORDRECHT / BOSTON / LONDON